Bioproducts

Bioproducts

T. Imanaka

Application of Recombinant DNA Technology to the Production of Useful Biomaterials

A. Kimura

Application of recDNA Techniques to the Production of ATP and Glutathione by the "Syntechno System"

D. Haferburg, R. Hommel, R. Claus, H.-P. Kleber

Extracellular Microbial Lipids as Biosurfactants

K. Soda, K. Yonaha

Applications of Stereoselectivity of Enzymes: Synthesis of Optically Active Amino and α-Hydroxy Acids, and Stereospecific Isotope-Labeling of Amino Acids, Amines and Coenzymes

W. K. Shieh, J. D. Keenan

Fluidized Bed Biofilm Reactor for Wastewater Treatment

THEORIA CUM PRAXI

AKADEMIE-VERLAG BERLIN

Bioproducts

Managing Editor: A. Fiechter

with 52 Figures and 45 Tables

Akademie-Verlag Berlin 1986

Die Originalausgabe erscheint im Springer-Verlag Berlin—Heidelberg—New York
als Volume 33 der Schriftenreihe
Advances in Biochemical Engineering/Biotechnology

Vertrieb ausschließlich für die DDR und die sozialistischen Länder Akademie-Verlag Berlin

ISBN 3-540-16380-8 Springer-Verlag Berlin Heidelberg New York Tokyo
ISBN 0-387-16380-8 Springer-Verlag New York Heidelberg Berlin Tokyo

ISBN 3-05-500205-9

Erschienen im Akademie-Verlag Berlin, DDR-1086 Berlin, Leipziger Straße 3—4
Lizenznummer: 202 · 100/547/86
Printed in the German Democratic Republic
Gesamtherstellung: VEB Druckerei „Thomas Müntzer", 5820 Bad Langensalza
LSV 1315
Bestellnummer: 763 643 6 (3070/33)
10800

Managing Editor

Editorial Board

Table of Contents

Application of Recombinant DNA Technology to the Production of Useful Biomaterials
T. Imanaka 1

Application of recDNA Techniques to the Production of ATP and Glutathione by the "Syntechno System"
A. Kimura 29

Extracellular Microbial Lipids as Biosurfactants
D. Haferburg, R. Hommel, R. Claus, H.-P. Kleber 53

Applications of Stereoselectivity of Enzymes: Synthesis of Optically Active Amino Acids and α-Hydroxy Acids, and Stereospecific Isotope-Labeling of Amino Acids, Amines and Coenzymes
K. Soda, K. Yonaha 95

Fluidized Bed Biofilm Reactors for Wastewater Treatment
W. K. Shieh, J. D. Keenan 131

Application of Recombinant DNA Technology to the Production of Useful Biomaterials

Tadayuki Imanaka
Department of Fermentation Technology, Faculty of Engineering, Osaka University, Yamada-oka, Suita-shi, Osaka 565, Japan

1 Introduction ... 2
2 Selection of Suitable Host-Vector Systems ... 2
2.1 Selection of Host Cells ... 2
2.2 Characteristics of Cloning Vectors ... 3
2.2.1 Plasmids ... 3
2.2.2 Phage Vectors ... 4
2.2.3 Cosmids ... 5
2.3 Transformation of Microorganisms ... 5
2.3.1 Transformation ... 5
2.3.2 Transduction ... 9
2.3.3 Conjugation ... 9
3 The Stability of Plasmids ... 9
3.1 Replication of Plasmid DNA and Its Partition to Daughter Cells ... 9
3.2 Instability of Recombinant Plasmids ... 10
3.3 Assessment of Stability of a Plasmid in Host Cells ... 11
3.4 Some Proposals to Ensure the Stability of Recombinant Plasmids ... 12
4 Improvement of Host-Vector Systems ... 14
4.1 Improvement of Vector ... 14
4.1.1 Control of Copy Number ... 14
4.1.2 Control of Gene Expression ... 14
4.2 Host-Vector System for Protein Secretion ... 17
4.3 Stability and/or Instability of Peptide Products ... 18
4.4 An Example of the Improvement of Host-Vector System ... 18
5 Examples of Genetically Engineered Microorganisms ... 20
6 Concluding Remarks ... 23
7 References ... 23

It is now possible to clone and express a gene of interest in widely differing hosts. Among many the steps necessary for the production of useful biomaterials in industry, gene cloning, scale-up of the process, and isolation and purification of products are the most important. For the efficient cloning of specific genes, suitable host-vector systems should be selected. For the successful cultivation of recombinant organisms in a large fermentor, both expression of cloned genes and stability of recombinant plasmid are essential. For the isolation of products, protein secretion systems must be improved.

The purpose of this review is to discuss some general concepts about the application of recombinant DNA technology in industry.

1 Introduction

In the bioindustry, mutagenesis has been used as a means for strain improvement of microorganisms. However, only a few nucleotide base pairs are changed by mutation. In contrast, a DNA fragment of more than 1000 base pairs can be manipulated through genetic engineering techniques. Accordingly, the use of recombinant DNA allows approaches in which useful genetic information can be inserted directly into microorganisms. Qualitative success might involve the cloning of foreign DNA such as human genes in microorganisms [1], and quantitative success includes the enhancement of products originating from the cloned genes due to the gene dosage effect [2,3]. Thus, plasmids may serve as powerful tools in producing peptides and/or metabolites in the cultivation of microorganisms.

Despite substantial "species barriers" such as nucleases, plasmid replication, transcription, translation, proteinases, and splicing barriers [4], a number of specific genes have been cloned in host cells. Unless cloned genes are fully expressed and are also kept in situ in vector plasmids during replication in coordination with the growth of host cells, it is not possible to employ the recombinant plasmid as an agent to enhance production of specific materials in industry. This review describes how to select a suitable host-vector system and also pays attention to the subject of expression and stability of cloned genes in the cultivation of host cells.

2 Selection of Suitable Host-Vector Systems

2.1 Selection of Host Cells

Despite the many examples of gene manipulation that are theoretically possible, some technical limitations for applications in industry still exist as follows; (1) Genetic maps do not always exist. (2) Gene exchange systems for industrially useful microorganisms, such as useful vectors and transformation procedures, are at an early stage of development. (3) Metabolic pathways leading from a raw material to the desired product such as antibiotics, are not made clear in many cases. Identification of all the steps would be necessary, and the number of genes involved for the conversion is a major limitation. Because of these limitations, a suitable host strain should be carefully selected to make good use of the characteristics of the host.

The most popular organisms and their characteristics as host cells are summarized below.

Escherichia coli: This is the best understood bacterium. More than 1000 genes have been identified [5]. Mandel and Higa [6] have shown that *E. coli* cells treated with cold calcium chloride become competent and it was subsequently shown that the competent cells could take up plasmid DNA as well as phage DNA.

Bacillus subtilis: This bacterium is genetically and biochemically well characterized. More than 300 genes have been identified on its circular chromosome [7]. *B. subtilis* is a non-pathogenic soil microorganism which grows strictly under aerobic conditions and, therefore, represents a safe host. *B. subtilis* does not contain pyrogenic lipopolysaccharides as does *E. coli*. *B. subtilis* is Gram-positive, and has a rather simple cell envelope structure which consists of a single layer of membrane. Therefore, secretory proteins such as amylase and protease are released directly into the culture medium [8]. This process obviates the necessity of disrupting cells and makes recovery and purification of secreted products simpler.

Bacillus stearothermophilus: This is a typical thermophile, and produces thermostable enzymes. The amount of cooling water required for a specific cultivation of a thermophile in a large scale fermentor would be reduced [9]. Since thermophilic bacteria can generally grow faster than mesophiles [10], the cultivation time would be shortened. Because of high cultivation temperature (> 55 °C), the number of possible contaminating organisms would be reduced.

Streptomyces spp.: Streptomyces species are well known as producers of several thousand antibiotics [11]. Many of them are constantly used in human and veterinary medicine and in agriculture. Application of recombinant DNA techniques to *Streptomyces* would play an important role in strain improvement aimed at increased antibiotic yields and the generation of novel antibiotics by incorporating parts of different natural antibiotic biosynthetic pathways into a strain.

Saccharomyces cerevisiae: This yeast is a commercially important strain, and is genetically well characterized [12]. Yeast is a suitable eukaryotic host, because it can be cultivated like bacteria on defined media and grows in colonies. The generation time is much shorter than those of other eukaryotic cells.

Animal cells: By and large, current methods of gene cloning in animal cells rely on the integration of the foreign DNA into the genome, using the genomes of special viruses (e.g. SV40) as cloning vehicles [13]. Attempts are being made to develop vectors which can be maintained in an extrachromosomal state. Mass culture of animal cells is an important method to be improved for the industrial applications.

Plant cells: Cloning in plant cells is currently being developed. It will open the way to the direct genetic modification of agricultural plants for the breeding. Plant viruses and the Ti plasmid of the bacterium *Agrobacterium tumefaciens* are usually used as vectors [14, 15].

2.2 Characteristics of Cloning Vectors

In principle, viral DNA and plasmid DNA molecules can be used as cloning vectors, because they are capable of replication in the host cells. However, since the characteristics of virus and plasmid are different in many points, we must be aware of their features, prior to the selection of suitable vector.

2.2.1 Plasmids

The plasmids of bacteria can be broadly divided into two classes. One class (the relaxed type of replication control) is characterized by a high copy number per chromosome. These plasmids are generally small and nonconjugative. Members of the second class (the stringent type of replication control) are much larger, and are maintained in low copy numbers. These plasmids generally are self-transmissible. However, this classification is not strict.

Criteria and/or the desirable characteristics for plasmid vector design are summarized as follows:

(1) A plasmid vector should be as small as possible, because the transformation efficiency of host cells decreases as the size of plasmid increases above 15 kilobases [16].

When extraneous genetic information is on a plasmid, the plasmid tends to be more unstable than the plasmid lacking the DNA region.
(2) Vectors should be replicated autonomously and be maintained stably in the desired host.
(3) The vector should possess the maximum number of unique restriction endonuclease cleavage sites (one cut for each enzyme if possible). Restriction site should not be located at DNA replication region.
(4) The vector should have a selectable marker to distinguish the transformant with the vector from nontransformed cells.
(5) It is desirable that a vector has an additional genetic marker which can be inactivated by the insertion of a foreign DNA fragment. The insertional inactivation is convenient to screen recombinant plasmids on the basis of altered phenotype.
(6) It is desirable that the copy number of vector plasmid is easily amplified by temperature shift or the addition of specific drug.
(7) It is desirable that the vector can be transferred in wide range of host species (shuttle vector).
(8) Secretion vector plasmid is desired to accumulate peptide product in a culture medium.
(9) It is also desirable that the regulation system of expression of cloned genes is set on the vector plasmid.

2.2.2 Phage Vectors

The most commonly used phage vectors are those derived from bacteriophage λ[15]. Phage λ is a temperate phage which can be virulent or lysogenic to the host cell, *E. coli*. For cloning experiments in *E. coli*, two types of λ vectors are available, i.e. insertion vector and replacement vector. The use of these phage vectors instead of plasmid vectors shows some advantages and disadvantages as follows:
(1) When the recombinant phage is lysogenized in host cells, the copy number per chromosome is unity. Therefore, the cloned gene is inherited stably with the chromosomal replication.
(2) Derivatives of λ that are smaller than 75% or larger than 105% of the size of wild-type λ are not packaged. Thus, the size of the cloned DNA fragment is limited.
(3) Recombinant DNA can be easily isolated from phage particles.
(4) Although a gene dosage effect is observed for recombinant plasmids, such an effect cannot be expected in case of a prophage because there exists only one copy per chromosome. When the phage growth is induced, lysis of host cell would be brought about.
(5) In the presence of high concentration of the phage head precursor and packaging proteins, recombinant λ DNA can be packaged in vitro. The packaged DNA can be introduced into *E. coli* cells with a 10 to 100 fold higher efficiency of transformation.
(6) Host strains are usually limited because of the high specificity of interaction between phage and its host cells.

Cloning into single-stranded phage vectors such as fd and M13 is desirable for special purposes, e.g. DNA sequencing and heteroduplex analysis.

Although phage ϱ11 was used for DNA cloning in *B. subtilis*, there is ample room for further improvement of the vector.

2.2.3 Cosmids

To join the advantages of both plasmid vector and phage vector, a new type of *E. coli* cloning vector "cosmid" was constructed [17]. Cosmid consists of a normal plasmid containing the *cos* site (cohesive ends) of phage λ, which promotes in vitro packaging. A small cosmid (<20 Md) is not efficiently packaged because of the small size. If the cosmid is joined to foreign DNA to increase the size and is a concatemeric form, it can be easily packaged. Therefore, only hybrid molecules carrying a large insert are recovered. After the introduction of cosmid into *E. coli*, the cosmid is maintained in the cell as a plasmid. Thus, cosmid is the ideal tool for constructing a gene bank.

2.3 Transformation of Microorganisms

The transfer of genetic material in microorganisms occurs either spontaneously in nature or experimentally in the laboratory. The processes can be classified in three categories, i.e., transformation, transduction and conjugation. General methods for the transfer of cloned genes to plant cells [14] and animal cells [13] have also been reported.

2.3.1 Transformation

Transformation was originally defined as genetic recombination in which naked DNA from one cell can enter and integrate into the chromosome of another cell [18]. Genetic transformation has been found not only in *Pneumococcus* but also in other bacterial genera, such as *Haemophilus*, *Streptococcus*, *Xanthomonas*, *Salmonella*, *Bacillus*, *Neisseria*, *Micrococcus*, and *Rhizobium*. When plasmid DNA is transferred into living cells and alters characteristics of the cells without genetic recombination, the phenomenon is also called as "transformation". Thus, the concept of transformation has been expanded. Two kinds of cells, competent cells and protoplasts (including spheroplasts and autoplasts), can be used as recipients. Table 1 shows the examples of transformation of microorganisms with plasmid DNA.

Transformation does not ordinarily occur in *E. coli*. However, it has been shown that *E. coli* cells treated with calcium chloride can take up naked DNA [6]. The reason for the competency might be explained by Ca^{++} partially removing the cell wall and/or periplasmic proteins and naked DNA is protected against deoxyribonuclease by combining with Ca^{++}. This $CaCl_2$ induced transformation procedure is generally applicable for Gram-negative bacteria. Other metal ions are more effective for yeast transformation [78].

Competent cells of *B. subtilis* can be transformed by either chromosomal DNA or plasmid DNA [48,82]. However, it must be mentioned that *B. licheniformis* competent cells can be transformed with chromosomal DNA but not with plasmid DNA [2]. Thus, applicability should be examined in each case.

To circumvent problems due to host specificity, a protoplast procedure has been used to transform many Gram-positive bacteria [51] and yeast [75]. The protoplast procedure consists of three main steps, i.e. (1) protoplast formation in hypertonic medium by cell-wall lytic enzyme such as lysozyme, (2) incorporation of plasmid

Table 1. Transformation of microorganisms with plasmid DNA

Microorganism	Procedure[a]	Remark	Ref.
Cyanobacteria			
Anacystis nidulans	C	Blue green algae, phototrophic prokaryotic organism. Heterocysts serve as sites of nitrogen fixation	19, 20)
Phototrophic bacteria			
Rhodopseudomonas sphaeroides	C	Photoorganotroph, facultatively aerobic, growing either anaerobically in the light or aerobically in the dark	21)
Gram-negative aerobic rods and cocci			
Agrobacterium tumefaciens	C[b]	Cause galls of plants	22, 23)
Azotobacter vinelandii	C	Capable of fixing molecular nitrogen	24)
Pseudomonas aeruginosa	C		25, 26)
Pseudomonas aureofaciens	C	Produce diffusible fluorescent pigment and exhibit antifungal activity	27)
Pseudomonas phaseolicola	C		28)
Pseudomonas putida	C	Produce diffusible fluorescent pigment	29, 30)
Rhizobium meliloti	C	Gall hypertrophies are produced on roots and stems of diverse plant species	31)
Gram-negative facultatively anaerobic rods			
Erwinia carotovora	C	Phytopathogen	32)
Escherichia coli	C	Best understood microorganism.	33)
Haemophilus influenzae	C	Strict parasites	34, 35)
Salmonella typhimurium	C	Ubiquitous and frequently the cause of infections in animals	36)
Gram-positive cocci			
Staphylococcus aureus	C	Hemolysins are produced.	37, 38)
Streptococcus lactis	P	Used by the dairy industry for the production of fermented milk products	39, 40)
Streptococcus pneumoniae	C	Lactic acid production, heterofermentative	41, 42)
Streptococcus sanguis	C	Lactic acid production, heterofermentative	43)
Streptococcus group F	C	Lactic acid production, heterofermentative	44)
Endospore-forming rods and cocci			
Bacillus amyloliquefaciens	P	Produce α-amylase	45)
Bacillus licheniformis	P	Produce penicillinase	2)
Bacillus megaterium	P		46)
Bacillus sphaericus	P		47)
Bacillus subtilis	C	Best understood in Gram-positive bacteria, enzyme secretion	48–50)
Bacillus subtilis	P		51)
Bacillus stearothermophilus	P	Thermophile, produce thermostable enzymes such as α-amylase and neutral protease	52)

Table 1. (continued)

Microorganism	Procedure[a]	Remark	Ref.
Bacillus thuringiensis	P	Used as a microbial insecticide, crystalline toxin produced upon sporulation	53)
Clostridium acetobutylicum	P	Strictly anaerobic, produce acetone and butanol	54)
Clostridium perfringens	A	Strictly anaerobic, produce acetic acid, butyric acid and butanol	55)
Actinomycetes and related organisms			
Brevibacterium flavum	P	Produce glutamate	56)
Corynebacterium glutamicum	P	Produce glutamate	56)
Corynebacterium herculis	P	Produce glutamate	56)
Microbacterium ammoniaphilum	P	Produce glutamate	56)
Streptomyces acrimycini	P	Exhibit anti-bacterial and anti-fungal activity	57)
Streptomyces ambofaciens	P	Produce macrolide antibiotic	58)
Streptomyces albus	P	Exhibit anti-bacterial activity	57)
Streptomyces azureus	P	Produce thiostrepton	57)
Streptomyces bikiniensis	P	Produce streptomycin	59)
Streptomyces chartreusis	P	Produce chartreusin	60)
Streptomyces coelicolor	P	Used in genetic studies, produce actinorhodin	61,62)
Streptomyces espinosus	P		63)
Streptomyces fradiae	P	Produce tylosin	57,64)
Streptomyces glaucescens	P	Exhibit anti-bacterial and anti-fungal activity	57)
Streptomyces griseofuscus	P	Appear to be nonrestricting, produce antibiotics	65)
Streptomyces griseus	P	Produce streptomycin	66)
Streptomyces kanamyceticus	P	Produce kanamycin	67)
Streptomyces kasugaensis	P	Produce kasugamycin	67)
Streptomyces lavendulae	P	Produce streptothricin	68)
Streptomyces lividans	P	Exhibit anti-bacterial activity	57,69)
Streptomyces parvullus	P	Produce actinomycin D	57,70)
Streptomyces pristinaespiralis	P		57)
Streptomyces rimosus	P	Produce oxytetracycline	67)
Streptomyces reticuli	P	Produce melanin pigment and leucomycin	71)
Streptomyces toyocaensis	P	Produce glycopeptide	64)
Streptomyces vinaceus	P	Produce vitamin B_{12}	67)
Streptomyces venezuelae	P	Produce chloramphenicol	67)
Yeast			
Kluyveromyces fragilis	C		72)
Kluyveromyces lactis	P	Transformed with linear killer plasmid DNA	73,74)
Saccharomyces cerevisiae	P	Best understood yeast, used in brewing and food industry	75–77)
Saccharomyces cerevisiae	C		78)
Schizosaccharomyces pombe	P	Fission yeast	79)
Fungi			
Neurospora crassa	P	Best understood fungus	80,81)

[a] C, competent cell; P, protoplast; A, autoplast

[b] Freeze and thaw

Table 2. Comparison of transformation procedures for *Bacillus subtilis* [83–87]

Procedure	Efficiency (transformants per µg DNA)		Advantages	Disadvantages
Competent cells	Unfractionated plasmid	2×10^4	Competent cells are readily prepared. Transformants can be selected readily on any medium. It takes relatively short time to get transformants	Requirement of plasmid oligomer makes shotgun cloning experiments difficult
	Linear	0		
	cccMonomer	4×10^1		
	ocMonomer	2×10^1		
	cccDimer	8×10^3		
	cccMultimer	2.6×10^5		
Plasmid rescue	Unfractionated plasmid	2×10^6	Oligomers are not required. Transformants can be obtained with linear DNA. Other features are the same as those for competent cells	Transformants contain both resident plasmid and incoming plasmid. These two kinds of plasmids should be separated by segregation or retransformation. Recipient cell must be Rec^+
Protoplasts	Unfractionated plasmid	3.8×10^6	Most efficient system. (up to 80% transformants in total regenerants). Competent cells are not required. Transformants can be obtained with linear DNA	Hypertonic and complex medium is required for the regeneration of cell wall. It takes relatively long time to get tranformants
	Linear	2×10^4		
	cccMonomer	3×10^6		
	cccDimer	2×10^6		
	cccMultimer	2×10^6		

ccc: covalently closed circular, oc: open circular

DNA into the protoplasts in the presence of polyethylene glycol, and (3) regeneration of cell wall.

Transformation by plasmid rescue is an alternative strategy for transforming *B. subtilis* [83,84]. If plasmid DNA is linearized by restriction endonuclease cleavage, no transformants of *B. subtilis* competent cells were obtained. However, if the recipient cell carries a homologous resident plasmid and if the restriction cut occurs within the homologous moiety, then the linear plasmid DNA can efficiently transform the host cell. In practice, foreign DNA is ligated to monomeric vector DNA and the recombinant molecules can be used to transform *B. subtilis* competent cells carrying a homologous plasmid.

Table 2 shows the comparison of different procedures of transforming *B. subtilis* as an example. It should be pointed out that deletion of a specific DNA region from plasmid sometimes occurs in the process of transformation [88]. Therefore, if necessary, the plasmid must be isolated from the transformants and examined.

2.3.2 Transduction

Transduction means the transfer of genetic material in microorganisms by means of viruses (phages). Therefore, it is also called as phage-mediated transformation. Phage vector can be used for gene cloning. Some phages mediate the interspecies transfer of plasmid DNA among the related bacteria through transduction process [89].

2.3.3 Conjugation

Many plasmids are conjugative. They have a cluster of genes for the transfer of themselves from one bacterium to another. Some conjugative plasmids can also transfer copies of chromosomal genes between bacteria. Such conjugative plasmids have been found in many strains of Gram-negative bacteria, but only in three Gram-positive genera, *Streptococcus*, *Streptomyces* and *Clostridium*.

Conjugative plasmids such as F plasmid and RP4 often transfer non-conjugative plasmids at about the same frequency as they transfer themselves [90]. A hybrid plasmid RP4::Mu has been used to create intergeneric recombinant strains for in vivo genetic engineering, e.g. RP4 is used for intergeneric gene transfer among wide range of host strains and bacteriophage Mu acts as a transposable genetic element [91,92].

3 The Stability of Plasmids

3.1 Replication of Plasmid DNA and Its Partition to Daughter Cells

Prior to the description of plasmid stability, the current concept of plasmid replication and its partition will be briefly mentioned. The mechanisms involved in the inheritance of plasmid have been of considerable interest. Three of the mechanisms seem to be crucial in this process; (1) a mechanism for the regulation of initiation of DNA replication ensures that the copy number of plasmid remains constant in dividing bacteria, (2) a mechanism for plasmid partition is thought to ensure their stable inheritance [90,93], and (3) another mechanism for coupling host cell division to plasmid proliferation prevents the loss of plasmid from host cells [94].

As predicted by the replicon hypothesis [95], most plasmids specify explicit functions that not only permit the plasmid to utilize the general replicative apparatus of the host cell but also determine the rate at which this utilization occurs. Extensive genetic analyses have revealed plasmid-encoded substances that are required for replication (positive effectors of replication) and others that act to keep the replication rate below the theoretical maximum (negative replication control substances) [96,97].

The presence of a mechanism for plasmid partition was suggested by the fact that low copy number plasmids (e.g. F plasmid of *E. coli*) were inherited stably into daughter cells during cell growth [95]. In fact, such genes that are essential for normal partition of the plasmid have been identified [23,98-100]. In the case of plasmid pSC101, a cis-acting small DNA segment is essential and sufficient for plasmid partition, however, no plasmid-encoded products were involved [101].

It has also been reported that the association of the origin of DNA replication with cytoplasmic membrane may serve to regulate in some way the initiation of DNA replication [102,103]. Taking advantage of this specific binding of a plasmid to membrane, the plasmid-free cells can be easily obtained from the plasmid carrying host cells since in the regeneration process of the protoplasts of plasmid carrier, normal cell division and partition of plasmid do not occur, resulting in the appearance of plasmid free cells at high frequencies [104].

The phenomenon of plasmid incompatibility, the inability of two related plasmids to coexist and be coinherited stably in dividing bacteria, can presumably result from interference with either mechanism. Studies of incompatibility have shown the involvement of the following; competition of partitioning mechanism, supression of plasmid replication by a common repressor [105], and inhibition of RNA primer formation by low molecular weight RNA followed by inhibition of initiation of plasmid replication [106].

Since many factors such as a specific DNA segment (e.g. origin of DNA replication), RNA, proteins originating from both chromosome of the host cell and plasmid, and cytoplasmic membrane, are involved in the replication and inheritance of plasmid, macroscopic rather than microscopic argument would be suitable to lead the general discussion on the plasmid stability.

3.2 Instability of Recombinant Plasmids

As mentioned above, plasmid is maintained under a subtle quasi-equilibrium condition in the host cell, and it is easily supposed that the recombinant plasmid carrying cloned genes would behave in a different manner from the original vector plasmid. Reasons for a general inclination towards instability of recombinant plasmids are summarized as follows;

(1) The higher the gene expression of a plasmid, the more segregants (plasmid free cells) tend to appear [107].

(2) Recombinant plasmid is quite unstable when the products of cloned genes are inhibitory to host cells [108].

(3) Phenotypic instability of plasmid is due either to the disappearance of entire plasmid from host cells or to the deletion of a specific region from the recombinant plasmid [107].

(4) Transposable elements such as transposon and insertion sequences promote deletion of DNA from recombinant plasmids [107,109].
(5) The instability of the plasmid-carrying cells can be explained by the fact that their growth rate is lower than that of host cells without plasmid and thus, by the enrichment of plasmid-free cells during the culture [4,107].
(6) Plasmids also exist which do not change the phenotype but the expression level decreases due to point mutation on the specific gene [110].
(7) The stability of plasmid cannot be guaranteed in chemostat culture, because plasmid loss from host cells is generally irreversible [4,110].

3.3 Assessment of Stability of a Plasmid in Host Cells

It is worthwhile examining the stability of plasmid in proliferating cells as a basis of genetic engineering. We, Imanaka and Aiba [4], considered that a mixed culture of plasmid-carrying (P) and plasmid-free (N) cells could be represented by the following reactions.

$$P \xrightarrow{\mu_1} (2-p)\,P$$
$$P \xrightarrow{\mu_1} N \qquad (N = pP) \tag{1}$$
$$N \xrightarrow{\mu_2} 2N$$

where p is the probability of plasmid loss and/or deletion per division of host cells, μ is specific growth rate, and subscripts 1 and 2 are cells with and without plasmid, respectively.

Under the assumption of exponential cell growth, we noted an important consequence for the mixed culture. With $N_0 = 0$ for the initial condition, fraction F of cells carrying plasmid in the total population decreases continuously with generation number n, according to the form

$$F_n = \frac{1 - \alpha - p}{1 - \alpha - p \cdot 2^{n(\alpha + p - 1)}} \tag{2}$$

where $\alpha = \mu_2/\mu_1$.

Taking n = 25 for instance, F_{25} vs. α is shown in Fig. 1, parameter being p. In general, α ranged nearly from 1.0 to 2.0. It is evident from Fig. 1 that F_{25} deteriorates drastically as α increases from 1.0 to 2.0. An extreme case, $\mu_2 = 0$ ($\alpha = 0$), may correspond to an auxotroph in a minimal medium; i.e., if plasmid, on which biosynthetic enzymes for nutrients required are coded, is lost, the cells cannot grow. Another case, $\mu_2 < 0$($\alpha < 0$), corresponds to the loss of drug-resistance plasmid. In other words, once the plasmid is lost, host cells are killed. It is clear from Eq. (2) that in batch system with any $\alpha > 1.0$ and $p > 0$, $F_\infty = 0$.

A typical series of steps in the scale-up is as follows: (1) inoculum from slant into conical flask (20 ml of liquid medium), (2) large flask (300 ml), (3) jar fermentor (10 l), (4) pilot fermentor (3,000 l), and (5) production size (100,000 l). If the inoculum size in each step is 3%, $n \cong 5$ is needed from step to step. Then the total amount of n required for the scale-up becomes n = 25. Consequently, if correlated with the produc-

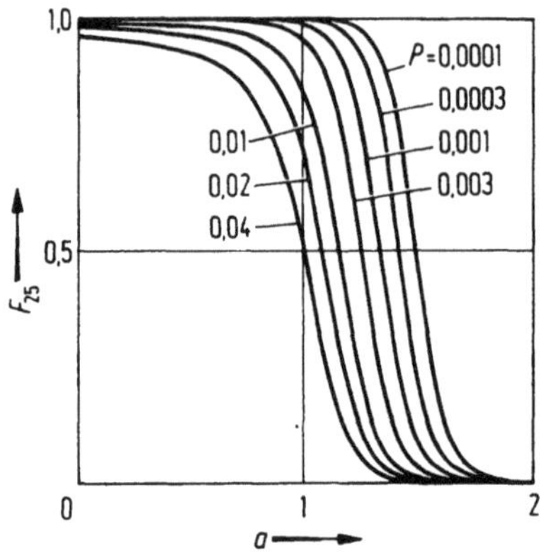

Fig. 1. F_{25} vs. α [4)]

tion performance, the F_{25} value might serve as a factor to judge whether or not the plasmid-carriers in question are of practical use in industry. It might be advisable to inoculate only a few cells (less than 50 cells per ml) into a fresh medium, because this small inoculum permits an assessment of F_{25}, i.e., the plasmid stability only in one step of the flask test.

In addition, batch reaction kinetics for product formation with recombinant cultures were further considered by other workers [111)].

3.4 Some Proposals to Ensure the Stability of Recombinant Plasmids

To ensure the stability of recombinant plasmid, it would be of crucial importance that the probability of plasmid loss p and α ($= \mu_2/\mu_1$) be lowered and the genetic structure remains unchanged. Some proposals are presented as follows;

(1) Plasmids can be stabilized by joining with a particular DNA segment that provides the partition function [98, 101, 112)].

(2) Plasmids can also be stabilized by joining with particular genes that couple host cell division to plasmid proliferation [94)].

(3) Environmental selective pressure is effective. For instance, only the cells carrying drug resistance plasmid can grow in the presence of a specific drug and the plasmid stability is secured.

(4) Runaway-replication plasmids increase their copy number after the cultures are shifted from a low to a high temperature. This temperature increase would decrease plasmid-absent cells, guaranteeing thus far the plasmid stability [113, 114)].

(5) It was confirmed experimentally that the higher the gene expression, the more segregants tended to appear. Conversely, plasmid is most likely to be kept unchanged when the gene expression is repressed. It might be advisable to use a thermoinducible expression system of plasmid vector and/or host chromosome. Cells are grown at normal temperatures to repress the gene expression in an early phase, followed by a temperature increase for the gene to be fully de-repressed [114)].

(6) Plasmids can be stabilized by joining with a particular DNA fragment or gene that improves growth rate of the host cells [115, 116)].

(7) Transposable elements promote insertion and deletion. The use of plasmid with such a transposable element should be avoided [107,109].
(8) It is recommendable to eliminate unnecessary DNA fragment from plasmids, because the redundant DNA becomes a burden on the host cell and also increases the probability of rearrangement of DNA in vivo.
(9) In the case of self-cloning, a recombination deficient mutant strain is required to avoid integration of a plasmid into the chromosome [117].
(10) Depending on the final products, a plasmid with optimal copy number should be used [2,3,118].
(11) Host cells should be improved so that a plasmid could become more stable [119].

In addition to the proposals, I would like to show some concrete examples of how plasmids were stabilized in host cells.

Antibiotics are usually added into culture media to screen plasmid carrier in small scale experiment. However, this procedure is not always practical in large scale reactor. Conversely, another method was developed by Miwa et al. [120] as follows: It has been known that an *E. coli* mutant strain requires streptomycin (Sm) for the cell growth [121]. Such a streptomycin dependent mutant strain (Sm^d) was firstly obtained from wild type *E. coli*, and was used as a host strain. Secondly, streptomycin independent gene (Sm^{id}), which is dominant to Sm^d gene, was cloned in pBR322. The recombinant plasmid (Sm^{id}) was quite stable in Sm^d host cells after successive cultures in the absence of Sm.

Likewise, another procedure was developed to exclude plasmid-free cells from the total population. Phage φ80, which carried temperature sensitive cI repressor gene, was lysogenized in *E. coli* host cell. A plasmid carrying normal cI gene was transferred in the host cells. When cultivation temperature was shifted up, the plasmid carrier could grow normally, although plasmid-free cells were lysed because of the thermal induction of phage φ80 [122].

pUB110 is a common vector plasmid in *B. subtilis*. Saito et al. [123] constructed the recombinant plasmid from pUB110 and a portion of the chromosomal gene. When *B. subtilis* was transformed with the plasmid, more than 40 copies of the plasmid were integrated into the host chromosome by Campbell type recombination. The plasmid existed also in cytoplasm at multi-copies. Thus, the plasmid sequences in both cytoplasm and chromosome were stably maintained.

Plasmid YRp7 [124] containing *ars* is fairly unstable in *Saccharomyces cerevisiae*. However, once a centromere region, CEN, carrying partition function is inserted into YRp7, the plasmid becomes stable [125].

Plasmid copy number is sometimes influenced by the cultivation conditions of host cells. High aeration rate and high cultivation temperature increased plasmid copy number, leading to plasmid stability, although the molecular mechanisms are not clear yet.

Immobilization of plasmid-carrying cells might be effective to maintain the plasmid stably in the reaction system.

4 Improvement of Host-Vector Systems

4.1 Improvement of Vector

4.1.1 Control of Copy Number

For gene cloning, a low copy number vector such as phage is desirable because of minimal stress to the host. However, multi-copy plasmids are in general necessary to realize the enhancement of the production of useful biomaterials. If the copy number of a vector could be artificially controlled, the vector would be more useful in industry. In fact, some vectors can be amplified by a temperature shift or by the addition of a drug [52,113,114,126]. It is also reported that the copy number of a plasmid can be changed by the addition or deletion of a specific DNA segment [127]. Thus, a suitable plasmid can be selected according to the purposes.

4.1.2 Control of Gene Expression

Control of gene expression is a very important point not only for plasmid stability but also for the increased productivity of useful biomaterials. The typical regulation systems of prokaryotic gene expression are summarized in Fig. 2. Improvement of each regulation system will be discussed item by item.

4.1.2.1 Improvement of Promoter

Unless the genes cloned in vector plasmids are fully expressed in host cells, it is difficult to expect the enhancement of productivity of specific materials. The DNA-dependent RNA polymerase recognizes specific DNA sequences, referred to as the promoter, where gene transcription is initiated [128] (Table 3). Most promoters of the *E. coli* chromosome so far sequenced share a consensus sequence. These domains are the —35 sequence and the Pribnow box in the —10 region [132,133]. Variation in the promoter sequence and the distance between —35 and —10 regions lead to the various levels of transcription for these genes [134,135]. Hybrid promoters that are

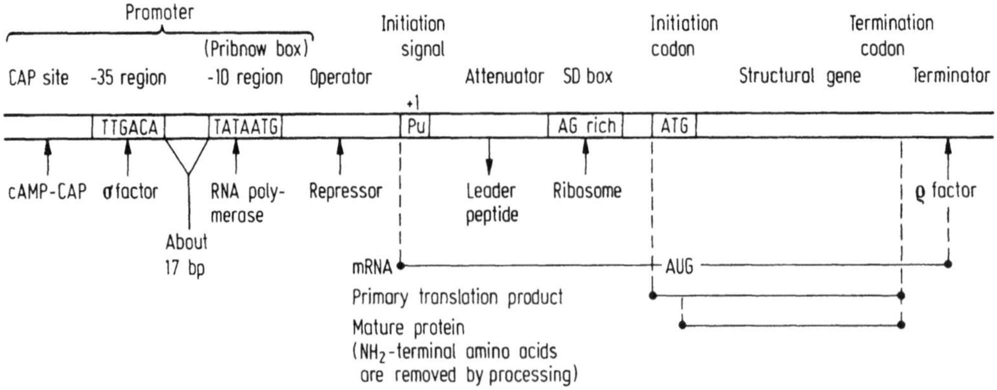

Fig. 2. Summary of prokaryotic gene expression system CAP: catabolite gene activator protein, cAMP: cyclic AMP, Pu: purine, SD box: Shine-Dalgarno sequence, ↑: binding, ↓: production

Table 3. Promoter sequences in bacteria [128–131]

Gene(s) and/or its product	−35 region	−10 region	σ factors
Most *E. coli* genes	TTGACA	TATAAT	σ^{70}
Heat-inducible genes of *E. coli*	TTGAAA	GATATA	σ^{32}
Many vegetative *B. subtilis* genes	TTGACA	TATAAT	σ^{55}
B. subtilis spoVG, spoVC and subtilisin	AGG-TT	GG-ATTG-T	σ^{37}
B. subtilis spoVG and *spoVC*	AAATC	TA-TG-TT-TA	σ^{32}
B. subtilis spoVG and *spoVC*	TT-AAA	CATATT	σ^{29}
B. subtilis genomic genes	CTAAA	CCGATAT	σ^{28}
Phage SPO1 for *B. subtilis*	T-AGGAGA--A	TTT-TTT	σ^{gp28}
Phage SPO1 for *B. subtilis*	CGTTAGA	GATATT	$\sigma^{gp33-34}$
B. subtilis spo0B	TTTTCT	TATAAT	unknown
B. stearothermophilus nprT	TTTTCC	TATTTT	unknown
B. thuringiensis crystal protein	AGTT-CA	$ATAA^{G}_{A}A$	unknown
B. megaterium protein C	CTAGTAACAA	CGCAAACAT	unknown

functional in *E. coli* have been constructed. These *tac* promoters were derived from sequences of the *trp* and the *lac*UV5 promoters [136, 137]. These promoters direct transcription more efficiently than either one of the parental promoters. Hybrid promoters can be repressed by the *lac* repressor and be de-repressed with the inducer, being useful for the controlled expression of foreign genes at high levels. In addition, the hybrid promoter is free from carbon catabolite repression even in the presence of glucose, because the CAP site (binding site of cAMP-CAP protein complex) of the *lac* operon is deleted.

When the *lac*UV5 promoter was inserted immediately after the constitutive *lpp* promoter, a cloned gene is not expressed in the absence of a *lac* inducer [138]. However, in the presence of an inducer, the cloned gene is transcribed from both the *lpp* and the *lac* promoters, which allow several times higher gene expression than the *lac*UV5 promoter alone.

It is also reported that the plasmid containing three successive *trp* promoters increases the expression level of a cloned gene in comparison with the case of one *trp* promoter [139].

There are several σ factors in *B. subtilis* which recognize the specific sequence of promoter, leading to the control of gene expression. Two overlapping promoters were cloned from *B. subtilis* chromosome and were found to be transcribed by *B. subtilis* σ^{55} and σ^{37} RNA polymerase holoenzymes during both growth and stationary phases [140].

Thus, the cloning of specific promoter and/or improvement of promoter are very important steps to set up the host-vector system.

4.1.2.2 Improvement of Gene Expression System other than Promoter

In negative control of gene expression, active repressor binds to operator, and blocks transcription by RNA polymerase. To start the transcription, either the addition of an inducer or the inactivation of repressor is required. Thermoinducible expression vectors have been constructed which carry the strong promoter such as P_L of bacteriophage λ [114]. The activity of this promoter is controlled by a temperature-sensitive

Table 4. Nucleotide sequences at the 3′ termini of small-subunit rRNAs [141]

Organism/source	Sequence
Eubacteria	
E. coli	***AACCUGCGGUUGGAUCACCUCCUUA$^{3'}$
B. stearothermophilus	***AAGGUGCGGCUGGAUCACCUCCUUUCUA
Archaebacteria	
Halobacterium halobium	***NNUCUGCGGCUGGAUCACCUCCU
Chloroplasts	
Zea mays	***AAGGUGCGGCUGGAUCACCUCCUUU
Euglena gracilis	***AAGGUGUGGCUGGAACAACUCCC
Mitochondria	
Yeasts	***AACCUGCGGUGGGCUUAUAAAUAUCUUAAAUAUUCUUACA
Wheat germ	***AACCUGUGGCUGGAUUGAAUC
Human	***AAAGUGCACUUGGACGAAC
Bovine	***AAAGUGUGCUUGGAUAAAU
Eukaryotic cells	
Mammals, frogs, silkworms, yeast	***AACCUGCGGAAGGAUCAUUA
Wheat germ	***AACCUGCGGAAGGAUCAUUG

repressor, product of the phage gene *cI*857. Since temperature is one of the easiest environmental factors to control during growth, thermoinducible gene expression system is useful.

For translation initiation in prokaryotes, the ribosome binding site, Shine-Dalgarno (SD) sequence, is important. SD sequence is recognized by the complementary nucleotide sequence of the 3′ terminus of small subunit rRNA (16S rRNA for *E. coli*) (Table 4). To enhance the translation efficiency, the SD sequence could be improved by the alteration of the sequence or the replacement with chemically synthesized sequence [142–144].

A transcription termination site, i.e. terminator, should also be considered for recombinant plasmids, because the deletion of terminator causes read-through of a gene to the following sequence, which may lead to instability of the mRNA [145,146].

Bacterial operons concerned with the biosynthesis of amino acids are often regulated by a process of attenuation [147]. If the specific regulation site, attenuator, were removed, the expression of the operon would be enhanced. In fact, such expression plasmid vectors which contain *E. coli* tryptophan promoter-operator region but lack the attenuator have been constructed [148].

DNA fragment synthesized chemically can be used to improve the regulation systems mentioned above, and also be used as a structural gene of desired peptide product. When the nucleotide sequence of a gene is designed, the amino acid codons used most frequently in the host cell should be selected to ensure efficient translation, since codon usage is fairly host-specific [149].

4.2 Host-Vector System for Protein Secretion

Secretion of protein is an essential function of all living cells. It is well known from a wealth of information that the secretion of a protein usually requires the presence of a relevant signal sequence [150,151]. In fact, many kinds of signal sequence have been discovered [152], and their structures and functions have been characterized [153]. Exploitation of secretion vectors has been attempted in order to increase the production of protein and to isolate products efficiently, because separation and purification of peptide products are the most costly processes in bioindustry.

In Gram-negative bacteria such as *E. coli*, the final destination of proteins passing through the plasma membrane is usually the periplasm. In the periplasmic-leaky (*lky*) mutant strains of *E. coli*, however, protein is accumulated in the culture medium during growth [154]. When the plasmid carrying the structural gene (*phoA*$^+$) of alkaline phosphatase was introduced into *lky* strain, the enzyme production was enhanced and 90% of total enzyme activity was released in the medium. Such a system might be very useful for the production and easier purification of peptide products.

Gram-positive bacteria such as *Bacillus* secrete many proteins directly into the culture medium. Now that some structural genes of extracellular enzymes (penicillinase, α-amylase, protease) have been cloned in *Bacillus* species [2,155–162], useful secretion vectors can be exploited. Indeed, *E. coli* β-lactamase was secreted from *B. subtilis* by the aid of α-amylase signal sequence [163]. Mammalian peptides such as interferon were also secreted by using these vectors [164]. The secretion rate and/or the total amount of protein accumulated in the medium is limited, e.g. not infinite. The number of specific sites for protein secretion was estimated as about 2×10^4

sites per cell [165]. It is also reported that increased secretion of α-amylase from *B. subtilis* caused by multiple copies of the cloned gene is not further increased by the chromosomal mutation enhancing the basic level of secretion [166]. Therefore, to further increase the secretion efficiency, structural genes for the secretion macinery should be cloned simultaneously in a host cell with the secretory enzyme gene.

4.3 Stability and/or Instability of Peptide Products

Talmadge and Gilbert [167] studied the biosynthesis of preproinsulin and proinsulin in *E. coli*. It is reported in the paper that the protein stability is affected by its cellular location in *E. coli*, i.e. molecules transported to the periplasm are more stable than those in the cytoplasm. This instability results from degradation of the molecules by some proteolytic activities.

It was also reported that peptide products (extracellular penicillinases) from the same gene in *B. subtilis* and *B. licheniformis* are processed differently by host-specific peptidases [168]. Such a host-specific processing of intracellular enzyme is reported for *E. coli* and *Neurospora crassa* [169]. Accordingly, it is important that any peptide product should be carefully examined when it is produced by different host strains, even if the product is expressed from the same structural gene.

If host cells secrete both proteases and peptide products from recombinant plasmid in the culture media, the products might be degraded by these proteases during growth. In this context, *B. subtilis* mutant strains lacking extracellular activities of both neutral protease and alkaline protease have been isolated [170] and would be useful for the accumulation of peptide products from a cloned gene.

Recent studies have revealed that the thermostability of a protein could be changed solely by a single amino acid substitution without any significant change of the three-dimensional protein structure [171,172]. Accordingly, we can enhance the thermostability of an enzyme by the site directed mutation of the cloned gene.

4.4 An Example of the Improvement of Host-Vector System

B. stearothermophilus is a thermophilic Gram-positive bacterium, which secretes a wide variety of proteins, and the physiological characteristics have been presented by many scientists [173,174].

We have isolated a drug resistance plasmid pTB19 (17.2 Md, Km^r Tc^r) from a thermophilic bacillus [175]. By using pTB19 and pUB110, we have also established a protoplast procedure for the transformation of *B. stearothermophilus* with plasmid DNA [52]. Many deletion plasmids were constructed from pTB19. Among them, pTB90, like the original plasmid pTB19, could transform both *B. subtilis* and *B. stearothermophilus*, whereas pTB51, pTB52 and pTB53 could replicate only in *B. subtilis*. The success or failure of transformation in both strains depended clearly on the species of plasmid DNA. Indeed, two replication determinants (RepA and RepB) were found on pTB19 [127].

RepA was functional only in *B. subtilis*, whereas RepB functioned in both *B. subtilis* and *B. stearothermophilus*. A deletion plasmid pTB90 carrying RepB contained the specific 1.0 Md *Eco*RI fragment of a cryptic plasmid pBSO2 from *B. stearothermophilus*. The presence of this 1.0 Md *Eco*RI fragment in various deletion plasmids

Table 5. Properties of plasmid pTB19 and its deletion derivatives [127]

Plasmid	Characteristics	Rep	1.0 Md *Eco*RI fragment	Copy no. in *B. subtilis*	Transformation of *B. stearothermophilus*	Copy no. in *B. stearothermophilus*
pTB19	Km^r Tc^r	A, B	—	1	+	1
pTB51	Km^r	A	—	8	—	—
pTB52	Tc^r	A	—	9	—	—
pTB53	Km^r Tc^r	A	—	8	—	—
pTB90	Km^r Tc^r	B	+	11	+	5 ~ 18[a]
pTB914	Km^r Tc^r	B	—	13	+[b]	43
pTB916	Km^r Tc^r	B	—	15	+[b]	—[c]
pTB913	Km^r	B	—	25	+[b]	39
pTB919	Km^r	B	+	13	+	5
pTB931	Km^r	B	+	14	+	4
pTB902	Tc^r	B	—	24	+[b]	60
pTB921	Tc^r	B	+	13	+	7

[a] Copy number, 5, in the absence of Tc; 18, in the presence of Tc
[b] Low transformation frequency
[c] Deletion plasmid appears frequently

increased transformation frequencies for *B. stearothermophilus* 10^3 to 10^4 times, and lowered copy numbers in the host strain to about one-tenth of those found for plasmids lacking this fragment (Table 5). Depending on the objectives, an optimal copy number plasmid can be selected from various plasmids (copy number 1 to 60 copies per chromosome).

The high frequency and efficiency of transformation in *B. stearothermophilus* protoplasts with pTB90 (6.7 Md, Km^r Tc^r; 20% transformants per regenerant and 2×10^7 transformants per μg of plasmid DNA, respectively) [52], compared with that in *B. subtilis* with plasmid pUB110 emphasize the usefulness of pTB90 as a vector plasmid for molecular cloning of specific genes in *B. stearothermophilus*. In fact, by using this system, penicillinase genes, *penP* and *penI*, of *B. licheniformis*, thermostable neutral protease gene, *nprT*, and thermostable α-amylase gene, *amyT*, from *B. stearothermophilus* have been cloned in the thermophile, and the enhancement of enzyme production was realized [156,158,160].

5 Examples of Genetically Engineered Microorganisms

The recent advances in the technology of molecular cloning and the development of expression vectors allow for example efficient production of clinically important mammalian peptides, or industrially useful enzymes by microorganisms carrying the appropriate genes [176]. Examples of peptide products and non-peptide products are shown in Table 6. Peptide can be directly synthesized as translation products by microorganisms. In contrast, non-peptide products are synthesized as the metabolites catalyzed by the specific enzymes originating from the cloned genes. For this reason, our work on amino acid production by a recombinant organism is briefly explained as an example.

For the purpose of studying the production of L-tryptophan by *E. coli*, the deletion mutants of the *trp* operon (*trpAE1*) were transformed with mutant plasmid (pSC101-trpI15 · 14) carrying the whole *trp* operon whose anthranilate synthase had been desensitized to tryptophan inhibition [3]. In addition, the deficiency of both tryptophan repressor (*trpR*) and tryptophanase (*tnaA*) was indispensable for host strains for the enhancement of L-tryptophan production (Table 7).

Table 6. Production of useful biomaterials by recombinant organisms

I. Peptide products
1. Hormone: insulin, growth hormone, somatostatin, endorphin, enkephalin, gastrin, calcitonin etc.
2. Physiologically active agent: interferon, interleukin, lymphokine, glutathione etc.
3. Vaccine: hepatitis B virus, foot-and-mouth disease virus, influenza virus, malaria etc.
4. Enzyme: amylase, protease, cellulase, pullulanase, urokinase, blood coagulation factor, rennin, cytochrome P-450, streptokinase, ribulosebisphosphate carboxylase etc.
5. Protein: collagen, serum albumin etc.

II. Non-peptide products
1. Amino acids: tryptophan, threonine, lysine, cystein, DOPA (3,4-dihydroxyphenylalanine) etc.
2. Vitamin: riboflavin etc.
3. Nucleoside, nucleotide: inosinic acid, guanylic acid etc.
4. Antibiotics: aminoglycoside, β-lactam etc.

The gene dosage effect on activities of specific enzymes of the *trp* operon was assessed. Assuming that tryptophan synthase (TSase) activity represents those of the *trp* operon enzymes, TSase activities were measured for high and low copy number plasmids (Table 8). The gene dosage effect was manifested in strain AE1 (*trpAE1*). Although the gene dosage effect could also be noted in strain Tna (*trpAE1 trpR tnaA*), the activity of TSase in Tna (pBR322-trp · I15) could not be measured because of the instability of plasmid.

We also examined the relationship between the copy number of the plasmids and tryptophan production (Table 8). The gene dosage effect observed and the limitation suggest the optimum plasmid copy number for tryptophan production.

It is interesting that tryptophan production was enhanced in proportion to the release from feedback inhibition by tryptophan. When the release became more advanced, tryptophan production was adversely affected. Therefore, the release from feedback inhibition for anthranilate synthase (ASase) activity should be optimized for tryptophan production.

The newly constructed strain *E. coli* W3110 *trpAE1 trpR tnaA* (pSC101-trpI15 · 14) exhibited high productivity of tryptophan ($6.2\ g\ l^{-1}\ 27\ h^{-1} = 0.23\ g\ l^{-1}\ h^{-1}$)[3] which exceeded any other data ever published. It must be emphasized that tetracycline

Table 7. Tryptophan production in shake flask cultures[a][3]

Strain[b]	Plasmid	Repression	Inhibition	Tryptophanase	Tryptophan (mg l^{-1})
AE1	pSC101-trp	+	+	+	7
AE1	pSC101-trp · I15	+	—	+	11
Ram	pSC101-trp	—	+	+	7
Ram	pSC101-trp · I15	—	—	+	70
Tna	pSC101-trp	—	+	—	8
Tna	pSC101-trp · I15	—	—	—	360

[a] Cells were grown in MTI medium (without anthranilic acid) at 37 °C for 36 h

[b] AE1, *E. coli* W3110 *trpAE1*; Ram, *E. coli* W3110 *trpAE1 trpR27*(Am); Tna, *E. coli* W3110 *trpAE1 trpR tnaA*

Table 8. Tryptophan synthase activities and tryptophan production[3]

Strain	Plasmid copy no. per chromosome	Tryptophan synthase (U mg^{-1} protein)	Tryptophan (g l^{-1})
AE1 (RP4-trp · I15)	1 ~ 3	4	
AE1 (pSC101-trp · I15)	~ 5	25	
AE1 (RSF1010-trp · I15)	10 ~ 50	47	
AE1 (pBR322-trp · I15)	60 ~ 80	114	
Tna (RP4-trp · I15)	1 ~ 3	36	1.7
Tna (pSC101-trp · I15)	~ 5	107	3.1
Tna (RSF1010-trp · I15)	10 ~ 50	215	2.6
Tna (pBR322-trp · I15)	60 ~ 80	—[a]	—

[a] —, Stable transformants difficult to obtain

Table 9. Characteristics of a recombinant microorganism for tryptophan production

Regulation	Host chromosome	Plasmid	Effect	Total effect
Repression	Deficiency of trp repressor (*trpR*)	—	Enhancement of *trp* operon enzymes	High accumulation of tryptophan
Inhibition	—	Insensitive to feedback inhibition (I^- for anthranilate synthase)	High rate of tryptophan synthesis	
Degradation of tryptophan	Deficiency of tryptophanase (*tnaA*)	—	No degradation of tryptophan	
—	Deletion of *trp* operon (*trpAE1*)	—	No recombination between chromosome and plasmid	Stable inheritance of plasmid
—	—	Tc^r	Screening of plasmid carrier in the presence of tetracycline	

was added in the medium to ensure the stability of the plasmid and that deletion mutants (*trpAE1*) were used as host strains to avoid the recombination of cloned *trp* operon with chromosomal DNA. These features are summarized in Table 9.

Although batch culture of above-mentioned strain was commendable in the production of tryptophan, the continuous culture was not successful in establishing steady state [110]. The failure of continuous culture was solely due to the instability of plasmid. Accordingly, the stability and/or instability of the plasmid becomes the most important factor concerning the actual use of a strain in industry.

6 Concluding Remarks

Genetic engineering originated in 1973/74 with the successful experiments performed by S. Cohen and his group in constructing recombinant plasmids [177,178]. Since then, many foreign DNA fragments have been cloned and transferred in many organisms including bacteria, yeast, fungi, plant cells and animal cells. The wide application of genetic engineering will induce the rise of a new type of industry as follows;

(1) Production of useful biomaterials (Table 6).

(2) Agricultural fields.
Increased yield of food and livestock, improvement of plant by cloning of nitrogen fixing enzyme genes, change of C3 plant to C4 plant, cloning resistance genes against damage by harmful insects, bioinsecticide such as *B. thuringiensis* etc.

(3) Prevention against environmental pollution.
Degradation of harmful materials by microorganisms etc.

(4) Others.
Energy production, microbial leaching in mining etc.

For these purposes, suitable upstream and downstream processes should be combined for the realization. In fact, downstream process such as separation and purification of end product are very costly and important for economy. If sugar-protein is a necessary product, sugar moiety must be added to the peptide product after the isolation from the culture broth. Thus, the specific modification process must also be developed. Although examples for plant cells and animal cells were not included in this review, the techniques of cell fusion and the cell culture would be improved and combined with recombinant DNA techniques in the future. The impact of this techniques for industry is enormous. However, the greatest benefits to the industry might well be eventually derived from the fundamental understanding of biological phenomena. Accordingly, basic knowledge of biology and the gene manipulation techniques can lead to the rational design of new useful biomaterials.

7 References

1. Derynck, R. et al.: Expression of human interferon-γ in heterologous systems, in: Experimental Manipulation of Gene Expression (ed. Inouye, M.), p. 247, New York, Academic Press 1983
2. Imanaka, T. et al.: J. Bacteriol. *147*, 776 (1981)
3. Aiba, S., Tsunekawa, H., Imanaka, T.: Appl. Environ. Microbiol. *43*, 289 (1982)
4. Imanaka, T., Aiba, S.: Ann. N.Y. Acad. Sci. *369*, 1 (1981)
5. Bachmann, B. J.: Microbiol. Rev. *47*, 180 (1983)

6. Mandel, M., Higa, A.: J. Mol. Biol. *53*, 159 (1970)
7. Henner, D. J., Hoch, J. A.: Microbiol. Rev. *44*, 57 (1980)
8. Priest, F. G.: Bacteriol. Rev. *41*, 711 (1977)
9. Aiba, S., Imanaka, T., Koizumi, J.: Ann. N.Y. Acad. Sci. *413*, 57 (1983)
10. Sonnleitner, B., Fiechter, A.: Trends Biotechnol. *1*, 74 (1983)
11. Hopwood, D. A.: Symp. Soc. Gen. Microbiol. *31*, 187 (1981)
12. Mortimer, R., Schild, D.: Microbiol. Rev. *44*, 519 (1980)
13. Spandidos, D. A., Wilkie, N. M.: Expression of exogenous DNA in mammalian cells, in: Transcription and Translation (eds. Hames, B. D., Higgins, S. J.), p. 1, Oxford, IRL Press 1984
14. Shaw, C. H. et al.: Gene *23*, 315 (1983)
15. Mainwaring, W. I. P. et al.: Nucleic Acid Biochemistry and Molecular Biology, p. 485, Oxford, Blackwell Scientific Publications 1982
16. Rodriguez, R. L., Tait, R. C.: Recombinant DNA Techniques: An Introduction, p. 4, Reading, Addison-Wesley Publishing Co. 1983
17. Collins, J., Hohn, B.: Proc. Natl. Acad. Sci. U.S.A. *75*, 4242 (1978)
18. Herskowitz, I. H.: Principles of Genetics, p. 182, New York, Macmillan Publishing Co. 1977[2]
19. Van den Hondel, C. A. M. J. J. et al.: Proc. Natl. Acad. Sci. U.S.A. *77*, 1570 (1980)
20. Golden, S. S., Sherman, L. A.: J. Bacteriol. *158*, 36 (1984)
21. Fornari, C. S., Kaplan, S.: ibid. *152*, 89 (1982)
22. Holsters, M. et al.: Mol. Gen. Genet. *163*, 181 (1978)
23. Gallie, D. R. et al.: J. Bacteriol. *157*, 739 (1984)
24. David, M., Tronchet, M., Dénarié, J.: ibid. *146*, 1154 (1981)
25. Sano, Y., Kageyama, M.: J. Gen. Appl. Microbiol. *23*, 183 (1977)
26. Mercer, A. A., Loutit, J. S.: J. Bacteriol. *140*, 37 (1979)
27. Aiba, S., Imanaka, T., Kondo, H.: unpublished data
28. Panopoulos, N. J., Staskawicz, B. J., Sandlin, D.: Search for plasmid associated traits and for a cloning vector in *Pseudomonas phaseolicola*, in: Plasmids of Medical, Environmental and Commercial Importance (eds. Timmis, K. N., Pühler, A.), p. 365, Amsterdam, Elsevier/North-Holland Biomedical Press 1979
29. Chakrabarty, A. M. et al.: Proc. Natl. Acad. Sci. U.S.A. *72*, 3647 (1975)
30. Bagdasarian, M. et al.: Gene *16*, 237 (1981)
31. Kiss, G. B., Kálmán, Z.: J. Bacteriol. *150*, 465 (1982)
32. Hinton, J. C. D. et al.: ibid. *161*, 786 (1985)
33. Cohen, S. N., Chang, A. C. Y., Hsu, L.: Proc. Natl. Acad. Sci. U.S.A. *69*, 2110 (1972)
34. Albritton, W. L., Bendler, J. W., Setlow, J. K.: J. Bacteriol. *145*, 1099 (1981)
35. Notani, N. K. et al.: ibid. *148*, 812 (1981)
36. Lederberg, E. M., Cohen, S. N.: ibid. *119*, 1072 (1974)
37. Lindberg, M., Sjostrom, J.-E., Johansson, T.: ibid. *109* 844 (1972)
38. Lindberg, M., Novick, R. P.: ibid. *115*, 139 (1973)
39. Kondo, J. K., McKay, L. L.: Appl. Environ. Microbiol. *43*, 1213 (1982)
40. Kondo, J. K., McKay, L. L.: ibid. *48*, 252 (1984)
41. Barany, F., Tomasz, A.: J. Bacteriol. *144*, 698 (1980)
42. Saunders, C. W., Guild, W. R.: Mol. Gen. Genet. *180*, 573 (1980)
43. LeBlanc, D. J., Hassell, F. P.: J. Bacteriol. *128*, 347 (1976)
44. LeBlanc, D. J., Cohen, L., Jensen, L.: J. Gen. Microbiol. *106*, 49 (1978)
45. Akamatsu, T., Sekiguchi, J.: Agric. Biol. Chem. *46*, 1617 (1982)
46. Brown, B. J., Carlton, B. C.: J. Bacteriol. *142*, 508 (1980)
47. McDonald, K. O., Burke, W. F. Jr.: J. Gen. Microbiol. *130*, 203 (1984)
48. Ehrlich, S. D.: Proc. Natl. Acad. Sci. U.S.A. *74*, 1680 (1977)
49. Gryczan, T. J., Contente, S., Dubnau, D.: J. Bacteriol. *134*, 318 (1978)
50. Keggins, K. M., Lovett, P. S., Duvall, E. J.: Proc. Natl. Acad. Sci. U.S.A. *75*, 1423 (1978)
51. Chang, S., Cohen, S. N.: Mol. Gen. Genet. *168*, 111 (1979)
52. Imanaka, T. et al.: J. Bacteriol. *149*, 824 (1982)
53. Martin, P. A. W., Lohr, J. R., Dean, D. H.: ibid. *145*, 980 (1981)
54. Reid, S. J. et al.: Appl. Environ. Microbiol. *45*, 305 (1983)
55. Heefner, D. L. et al.: J. Bacteriol. *159*, 460 (1984)
56. Katsumata, R. et al.: ibid. *159*, 306 (1984)
57. Kieser, T. et al.: Mol. Gen. Genet. *185*, 223 (1982)

58. Richardson, M. A. et al.: Gene *20*, 451 (1982)
59. Horinouchi, S., Kumada, Y., Beppu, T.: J. Bacteriol. *158*, 481 (1984)
60. Murakami, T. et al.: J. Antibiot. *36*, 1305 (1983)
61. Bibb, M. J., Ward, J. M., Hopwood, D. A.: Nature *274*, 398 (1978)
62. Malpartida, F., Hopwood, D. A.: ibid. *309*, 462 (1984)
63. Manis, J. J., Highlander, S. K.: Gene *18*, 13 (1982)
64. Jones, M. D. et al.: Plasmid *11*, 92 (1984)
65. Larson, J. L., Hershberger, C. L.: J. Bacteriol. *157*, 314 (1984)
66. Ohnuki, T., Imanaka, T., Aiba, S.: Gene *25*, 155 (1983)
67. Nabeshima, S. et al.: J. Antibiot. *37*, 1026 (1984)
68. Kobayashi, T. et al.: ibid. *37*, 368 (1984)
69. Thompson, C. J., Ward, J. M., Hopwood, D. A.: Nature *286*, 525 (1980)
70. Bibb, M., Schottel, J. L., Cohen, S. N.: ibid. *284*, 526 (1980)
71. Schrempf, H.: J. Bacteriol. *151*, 701 (1982)
72. Das, S., Kellermann, E., Hollenberg, C. P.: ibid. *158* 1165 (1984)
73. Das, S., Hollenberg, C. P.: Curr. Genet. *6*, 123 (1982)
74. De Louvencourt, L. et al.: J. Bacteriol. *154*, 737 (1983)
75. Beggs, J. D.: Nature *275*, 104 (1978)
76. Hinnen, A., Hicks, J. B., Fink, G. R.: Proc. Natl. Acad. Sci. U.S.A. *75*, 1929 (1979)
77. Struhl, K. et al.: ibid. *76*, 1035 (1979)
78. Ito, H. et al.: J. Bacteriol. *153*, 163 (1983)
79. Beach, D., Nurse, P.: Nature *290*, 140 (1981)
80. Case, M. E. et al.: Proc. Natl. Acad. Sci. U.S.A. *76*, 5259 (1979)
81. Hughes, K. et al.: ibid. *80*, 1053 (1983)
82. Anagnostopoulos, C., Spizizen, J.: J. Bacteriol. *81*, 741 (1961)
83. Gryczan, T., Contente, S., Dubnau, D.: Mol. Gen. Genet. *177*, 459 (1980)
84. Weinrauch, Y., Dubnau, D.: J. Bacteriol. *154*, 1077 (1983)
85. Old, R. W., Primrose, S. B.: Principles of Gene Manipulation, p. 48, Oxford, Balckwell Scientific Publications 1981[2]
86. Mottes, M. et al.: Mol. Gen. Genet. *174*, 281 (1979)
87. Canosi, U., Iglesias, A., Trautner, T. A.: ibid. *181*, 434 (1981)
88. Clancy, S. et al.: J. Bacteriol. *159*, 1065 (1984)
89. Ruhfel, R. E., Robillard, N. J., Thorne, C. B.: ibid. *157*, 708 (1984)
90. Hardy, K.: Bacterial Plasmids, p. 21, American Society for Microbiology 1981
91. Murooka, Y., Takizawa, N., Harada, T.: J. Bacteriol. *145*, 358 (1981)
92. Murooka, Y. et al.: Agric. Biol. Chem. *47*, 1807 (1983)
93. Scott, J. R.: Microbiol. Rev. *48*, 1 (1984)
94. Ogura, T., Hiraga, S.: Proc. Natl. Acad. Sci. U.S.A. *80*, 4784 (1983)
95. Jacob, F., Brenner, S., Cuzin, F.: Cold Spring Harbor Symp. Quant. Biol. *28*, 329 (1963)
96. Nordström, K., Ingram, L. C., Lumbäck, A.: J. Bacteriol. *110*, 562 (1972)
97. Novick, R. P.: Mol. Gen. Genet. *135*, 131 (1974)
98. Ogura, T., Hiraga, S.: Cell *32*, 351 (1983)
99. Miller, C. A. et al.: Gene *24*, 309 (1983)
100. Austin, S. J.: J. Bacteriol. *158*, 742 (1984)
101. Meacock, P. A., Cohen, S. N.: Cell *20*, 529 (1980)
102. Sparks, R. B., Helinski, D. R.: Nature *277*, 572 (1979)
103. Winston, S., Sueoka, N.: Proc. Natl. Acad. Sci. U.S.A. *77*, 2834 (1980)
104. Novick, R. et al.: Plasmid *3*, 348 (1980)
105. Timmis, K. N. et al.: Cold Spring Harbor Symp. Quant. Biol. *43*, 105 (1979)
106. Tomizawa, J., Itoh, T.: Proc. Natl. Acad. Sci. U.S.A. *78*, 6096 (1981)
107. Imanaka, T., Tsunekawa, H., Aiba, S.: J. Gen. Microbiol. *118*, 253 (1980)
108. Von Meyenburg, K. et al.: Cold Spring Harbor Symp. Quant. Biol. *43*, 121 (1979)
109. Tsunekawa, H. et al.: J. Gen. Microbiol. *127*, 93 (1981)
110. Dwivedi, C. P., Imanaka, T., Aiba, S.: Biotechnol. Bioeng. *24*, 1465 (1982)
111. Ollis, D. F., Chang, H. T.: ibid. *24*, 2583 (1982)
112. Skogman, G., Nilsson, J., Gustafsson, P.: Gene *23*, 105 (1983)
113. Uhlin, B. E. et al.: ibid. *6*, 91 (1979)
114. Remaut, E., Tsao, H., Fiers, W.: ibid. *22*, 103 (1983)

115. Edlin, G., Tait, R. C., Rodriguez, R. L.: Biotechnol. *2*, 251 (1984)
116. Biel, S. W., Hartl, D. L.: Genetics *103*, 581 (1983)
117. Tanaka, T.: Mol. Gen. Genet. *175*, 235 (1979)
118. Bremer, E. et al.: ibid. *179*, 13 (1980)
119. Ostroff, G. R., Péne, J. J.: J. Bacteriol. *156*, 934 (1983)
120. Miwa, K., Nakamori, S., Momose, H.: Seikagaku *53*, 816 (1981)
121. Momose, H., Gorini, L.: Genetics *67*, 19 (1971)
122. Sumitani, T. et al.: Japan patent 58-78589 (1983)
123. Saito, H., Anzai, H., Kawamura, F.: Multicopy intergration vectors in *Bacillus subtilis*, in: Genetics of Industrial Microorganisms (eds. Ikeda, Y., Beppu, T.), p. 125, Tokyo, Kodansha 1982
124. Struhl, K. et al.: Proc. Natl. Acad. Sci. U.S.A. *76*, 1035 (1979)
125. Clarke, L., Carbon, J.: Nature *287*, 504 (1980)
126. Close, T. J., Zaitlin, D., Kado, C. I.: Plasmid *12*, 111 (1984)
127. Imanaka, T. et al.: J. Gen. Microbiol. *130*, 1399 (1984)
128. Rosenberg, M., Court, D.: Ann. Rev. Genet. *13*, 319 (1979)
129. Taylor, W. E. et al.: Cell *38*, 371 (1984)
130. Grossman, A. D., Erickson, J. W., Gross, C. A.: ibid. *38*, 383 (1984)
131. Takagi, M., Imanaka, T., Aiba, S.: J. Bacteriol. *163*, 824 (1985)
132. Pribnow, D.: Proc. Natl. Acad. Sci. U.S.A. *72*, 784 (1975)
133. Gilbert, W.: Starting and stopping sequences for the RNA polymerase, in: RNA Polymerase (eds. Losick, R., Chamberlin, M.), p. 193, New York, Cold Spring Harbor Laboratory 1976
134. Stefano, J. E., Gralla, J. D.: Proc. Natl. Acad. Sci. U.S.A. *79*, 1069 (1982)
135. Russell, D. R., Bennett, G. N.: Gene *20*, 231 (1982)
136. De Boer, H. A., Comstock, L. J., Vasser, M.: Proc. Natl. Acad. Sci. U.S.A. *80*, 21 (1983)
137. Amann, E., Brosius, J., Ptashne, M.: Gene *25*, 167 (1983)
138. Nakamura, K., Inouye, M.: EMBO J. *1*, 771 (1982)
139. Goeddel, D. V. et al.: Nucleic Acids Res. *8*, 4057 (1980)
140. Wang, P.-Z., Doi, R. H.: J. Biol. Chem. *259*, 8619 (1984)
141. Kozak, M.: Microbiol. Rev. *47*, 1 (1983)
142. Gold, L. et al.: Ann. Rev. Microbiol. *35*, 365 (1981)
143. Gheysen, D. et al.: Gene *17*, 55 (1982)
144. Jay, E. et al.: Nucleic Acids Res. *10*, 6319 (1982)
145. Lee, N., Nakamura, K., Inouye, M.: J. Bacteriol. *146*, 861 (1981)
146. Gentz, R. et al.: Proc. Natl. Acad. Sci. U.S.A. *78*, 4936 (1981)
147. Yanofsky, C.: Nature *289*, 751 (1981)
148. Tacon, W. C. A. et al.: Gene *23*, 255 (1983)
149. Ikemura, T., Ozeki, H.: Cold Spring Harbor Symp. Quant. Biol. *47*, 1087 (1983)
150. Blobel, G.: Proc. Natl. Acad. Sci. U.S.A. *77*, 1496 (1980)
151. Kreil, G.: Ann. Rev. Biochem. *50*, 317 (1981)
152. Watson, M. E. E.: Nucleic Acids Res. *12*, 5145 (1984)
153. Silhavy, T. J., Benson, S. A., Emr, S. D.: Microbiol. Rev. *47*, 313 (1983)
154. Lazzaroni, J. C., Portalier, R.: Eur. J. Appl. Microbiol. Biotechnol. *16*, 146 (1982)
155. Gray, O., Chang, S.: J. Bacteriol. *145*, 422 (1981)
156. Fujii, M., Imanaka, T., Aiba, S.: J. Gen. Microbiol. *128*, 2997 (1982)
157. Palva, I.: Gene *19*, 81 (1982)
158. Aiba, S., Kitai, K., Imanaka, T.: Appl. Environ. Microbiol. *46*, 1059 (1983)
159. Yamazaki, H. et al.: J. Bacteriol. *156*, 327 (1983)
160. Fujii, M. et al.: ibid. *154*, 831 (1983)
161. Vasantha, N. et al.: ibid. *159*, 811 (1984)
162. Yang, M. Y., Ferrari, E., Henner, D. J.: ibid. *160*, 15 (1984)
163. Palva, I. et al.: Proc. Natl. Acad. Sci. U.S.A. *79*, 5582 (1982)
164. Palva, I. et al.: Gene *22*, 229 (1983)
165. Ito, K., Bassford, P. J., Beckwith, J.: Cell *24*, 707 (1981)
166. Sibakov, M., Sarvas, M., Palva, I.: FEMS Microbiol. Letters *17*, 81 (1983)
167. Talmadge, K., Gilbert, W.: Proc. Natl. Acad. Sci. U.S.A. *79*, 1830 (1982)
168. Imanaka, T. et al.: J. Gen. Microbiol. *129*, 2621 (1983)
169. Alton, N. K. et al.: Proc. Natl. Acad. Sci. U.S.A. *79*, 1955 (1982)
170. Kawamura, F., Doi, R. H.: J. Bacteriol. *160*, 442 (1984)

171. Yutani, K. et al.: Nature (London) *267*, 274 (1977)
172. Matsumura, M. et al.: J. Bacteriol. *160*, 413 (1984)
173. Sonnleitner, B., Fiechter, A.: Trends Biotechnol. *1*, 74 (1983)
174. Ljungdahl, L. G.: Adv. Microb. Physiol. *19*, 149 (1979)
175. Imanaka, T., Fujii, M., Aiba, S.: J. Bacteriol. *146*, 1091 (1981)
176. Office of Technology Assessment: Genetic Technology, Boulder, Westview Press 1982
177. Cohen, S. N. et al.: Proc. Natl. Acad. Sci. U.S.A. *70*, 3240 (1973)
178. Morrow, J. F. et al.: ibid. *71*, 1743 (1974)

Application of recDNA Techniques to the Production of ATP and Glutathione by the "Syntechno System"

Akira Kimura
Research Institute for Food Science, Kyoto University, Uji, Kyoto 611, Japan

1 Construction of a Microbial Bioreactor (or Biocatalyst) System to Produce Various Useful Compounds ... 30
2 Improving the ATP Regenerating Process by recDNA Techniques ... 31
2.1 Cloning of the Glucokinase Gene (*glk*) ... 32
2.2 Phosphorylation of AMP to ATP by Dried Cells of Genetically Engineered *E. coli* Harboring the *glk* Gene ... 34
2.2.1 Effect of Glucokinase Activity on ATP Production ... 34
2.3 Cloning of the Phosphofructokinase (*pfk*) and Triosephosphateisomerase (*tpi*) Genes 35
2.3.1 Effect of Plasmid pLC16-4 on ATP Formation ... 35
2.3.2 Subcloning of pLC16-4 (Separation of *pfk* and *tpi* Genes) ... 36
2.3.3 Gene Dosage Effects of *pfk* and *tpi* on Enzyme Activity ... 38
2.3.4 Effect of *pfk* and *tpi* on ATP Formation ... 39
3 Production of Glutathione by the "Syntechno System" ... 39
3.1 Production Strategy ... 39
3.2 Cloning of a Gene (*ghs*-I) Responsible for GSH-I Activity ... 41
3.2.1 Glutathione-deficient Mutants ... 41
3.2.2 Susceptibility of Mutants to Metal Compounds ... 41
3.2.3 Isolation of *gsh*-I with Various Restriction Endonucleases ... 42
3.2.4 Characterization and Analysis of *gsh*-I ... 42
3.2.5 Accumulation of the Dipeptide intermediate γ-Glutamylcysteine by the Cells Carrying the Cloned *gsh*-I Gene ... 43
3.3 Cloning of the *gsh*-II Gene for GSH-II Activity ... 43
3.3.1 Isolation of *gsh*-II in an Effort to Construct a More Efficient Glutathione Producer 43
3.3.2 Characterization and Analysis of *gsh*-II ... 44
3.3.3 Subcloning and Stabilization of *gsh*-II ... 44
3.3.4 DNA Base Sequence of *gsh*-II ... 46
3.4 Construction of a Hybrid Plasmid Containing both *gsh*-I and II in Various Ratios ... 46
3.4.1 Construction of a Hybrid Plasmid pGS500 Containing *gsh*-I and *gsh*-II ... 46
3.4.2 Construction of a Hybrid Plasmid Containing *gsh*-I and -II in Various Ratios ... 48
3.5 Production of Glutathione by the "Syntechno System" with Genetically Engineered Cells 48
4 Future Aspects (Protein Engineering) ... 50
5 References ... 50

Recent new techniques such as genetic engineering and/or cell fusion have made it possible to manipulate genes and to make microorganisms intentionally have various desired properties. Using these useful techniques, we have succeeded in the breeding of functional microorganisms which can produce large amounts of ATP or glutathione (a detoxification drug for liver). This review deals with the construction of microbial bioreactor or biocatalyst systems by the use of recDNA techniques and with the production of glutathione by genetically engineered cells.

1 Construction of Microbial Bioreactor (or Biocatalyst) System to Produce Various Useful Compounds

By the use of dried cells of yeasts or *Escherichia coli*, microbial bioreactor or biocatalyst systems have been constructed which produce various useful compounds [1]. As shown in Fig. 1, such a biorector consists of two stages: (1) an energy (ATP)-regenerating process and (2) a substrate-converting process. In the former, ATP is regenerated efficiently, and this ATP is then used in the latter to produce various useful compounds. The mechanism of this biorector system has been analyzed from various standpoints. It was shown by the use of various respiratory-deficient mutants of *Saccharomyces cerevisiae* [2] that ATP was regenerated through the glycolytic pathway. This system required a high amount of inorganic phosphate, which at the beginning of the reaction inhibited the activity of hexokinase, the first enzyme of the gly-

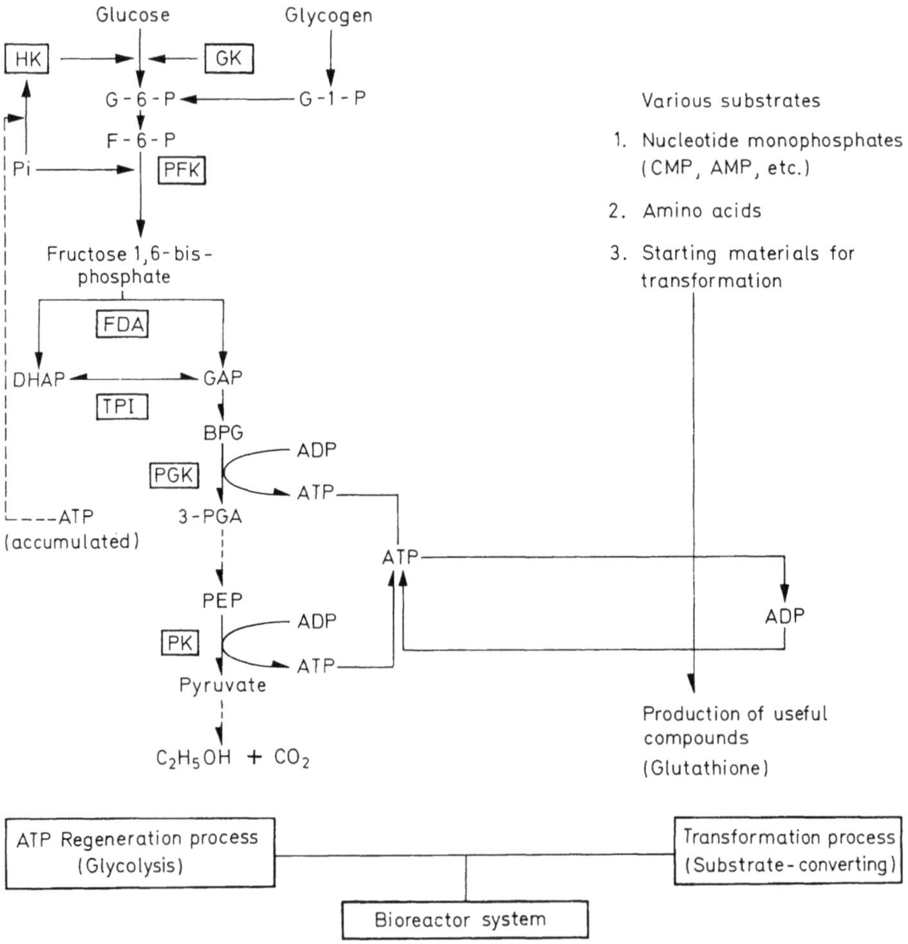

Fig. 1. Machinery of the bioreactor or biocatalyst system

colytic pathway [3]. Therefore, glucose was not used as an energy source until some amount of ATP was accumulated in the reaction mixture. At the beginning, ATP was produced by the use of glycogen present in cells, because glycogen phosphorylase, which was not inhibited by the high amount of phosphate, hydrolyzes glycogen gradually to produce glucose-6-phosphate (G6P). The G6P thus generated flowed into the glycolytic pathway and was then further metabolized to generate ATP. When the ATP level was increased, competitive inhibition of hexokinase by phosphate was overcome so that more glucose was phosphorylated and further metabolized, and then more ATP was regenerated. Thus, the ATP regeneration system operated automatically and at an increasing rate [3]. Therefore, the amount of glycogen present in the cells plays an important role at the beginning of the reaction, and cells without glycogen cannot carry out the reaction. These mechanisms were revealed by the use of a petite-negative yeast, *Hansenula jadinii* [3,4,5,6]. Aerobically cultured cells of *H. jadinii*, which did not contain glycogen, were unable to generate ATP and thus the inhibitory effect of phosphate on hexokinase was not relieved. On the other hand, anaerobically cultured cells of *H. jadinii* [4,5] containing a large amount of glycogen could carry out the reaction very efficiently [3]. This mechanism was confirmed by computer simulation [7]. The calculated time-dependent concentrations of glucose, FBP, adenosine and its nucleotides (AMP and ATP) were in good agreement with experimental values qualitatively as well as quantitatively.

By the use of this microbial (yeast) biorector or biocatalyst system, various useful compounds have been formed, such as cytidine-diphosphate-choline and sugar nucleotides. To introduce recDNA techniques to this system, we, at first, tried to construct an effective bacterial (*E. coli*) system as a model, and then apply it to the yeast system as discussed elsewhere.

2 Improving the ATP Regenerating Process by recDNA Techniques

To produce useful compounds efficiently by the microbial catalyst system, the ATP generating process mentioned above should be coupled with various substrate-converting process, depending on what kinds of products are envisioned. However, success of the overall process will depend upon the development of an efficient ATP generating process. Therefore, as a common process, the ATP regenerating process should be made efficient.

Initial experiments with dried *E. coli* cells indicated that the glucose-phosphorylating activity was very easily destroyed, so glucose could not be used as an energy source. *E. coli* transports glucose by the phosphoenolpyruvate phosphotransferase (PTS) system, which is readily inactivated in dried *E. coli* cells. To improve this situation, we tried to clone the glucokinase gene [8,9] from *E. coli*, and introduce it to *E. coli* cells. Recently, we have also succeeded in cloning a gene for yeast glucokinase [10]. In this Section, improvement of the ATP regenerating process will be described, and thereafter, the production of glutathione by gene-engineered cells of *E. coli* will be discussed as an example of improving the substrate-converting process.

At first, *E. coli* cells were employed as a model to construct a functional genetically engineered strain having genes of glycolytic enzymes: GLK (glucokinase), PFK (phosphofructokinase), and TPI (triosephosphate isomerase).

2.1 Cloning of the Glucokinase Gene (glk) [8-10]

Glucose, as an energy source, can be metabolized easily through glycolysis by almost any microorganism. The first step is the phosphorylation of glucose to glucose-6-phosphate (G6P). However, the phosphorylation is carried out in slightly different ways, depending on the organism. In the yeast *Saccharomyces cerevisiae* there are three glucose phosphorylating enzymes (hexokinase I, hexokinase II, and glucokinase), which are involved in glucose phosphorylation. On the other hand, *E. coli* is usually believed to phosphorylate glucose by coupling with its membrane transportation through the phosphotransferase system (PTS), although the presence of glucose phosphorylating enzymes have been reported [11]. The PTS system seemed very labile and easily destroyed when cells were dried. Since dried cells of *E. coli* were unable to phosphorylate glucose, glucose-6-phosphate (G6P) was employed as the energy source in the *E. coli* biocatalyst system. However, since the use of G6P instead of glucose was quite uneconomical, we tried to construct *E. coli* cells which could utilize glucose in biocatalyst system.

E. coli has three glucose phosphorylating enzymes. They are ATP-dependent glucokinase (GLK), glucose phosphotransferase (GPT), and mannose phosphotransferase (MPT). GPT is highly specific for glucose. MPT is active on glucose, mannose, and several other sugars. These enzymes have been thought to be of major importance in permitting rapid growth on glucose and mannose. Almost all studies of glucose metabolism in *E. coli* have been carried out on GPT and MPT and little information on GLK is available.

An *E. coli* mutant deficient in GPT, MPT, and GLK activities has been isolated and reported to be unable to grow on glucose, but a second mutant lacking only GPT and MPT was able to grow on glucose [12]. This indicated that GLK could function for glucose catabolism in *E. coli* cells, and suggested that introducing the *glk* gene might be of use to improve *E. coli* cells.

For the cloning of a gene responsible for GLK, *E. coli* chromosomal DNA was digested with restriction endonuclease *Hin*dIII, and the resultant DNA fragments were ligated with the plasmid vector pBR322 which also had been digested with *Hin*dIII. *E. coli* ZSC 112L (*gpt mpt glk*) having a decreased level of GLK activity was transformed with the whole ligation mixture prepared above. The transformants were selected as ampicillin-resistant, very light pink colonies on glucose-MacConkey, white colonies on mannose-MacConkey, and blue ones on Davis-Mingioli minimal medium [8, 13]. These results indicated that the cloned gene was for GLK, but not for MPT or GPT, the latter two enzymes show different colors on these media [12].

The hybrid plasmid in one of the transformants was isolated and designated pGK100. The circular restriction map of pGK100 was determined as shown in Fig. 2. This hybrid plasmid was 8.8 megadaltons (MDa) and had an insert of 6.0 MDa of a chromosomal DNA fragment of *E. coli* in the *Hin*dIII site of the vector. The cloned fragment had four susceptible sites for each of *Eco*RI, *Pst*I, and *Mlu*I (Fig. 2) [8].

To locate the *glk* gene on this insert, pGK100 was completely digested with *Mlu*I. After electrophoresis of the digestion mixture on agarose gels, the DNA fragments were extracted. Four kinds of linear DNA fragments (F1, F2, F3, and F4) of 1.2, 0.82, 0.5, and 6.2 MDa, respectively, were obtained (Fig. 2a). Each fragment (F1, F2, and F3) was ligated with F4 by T_4 DNA ligase, and F4 was self-ligated. The ligation

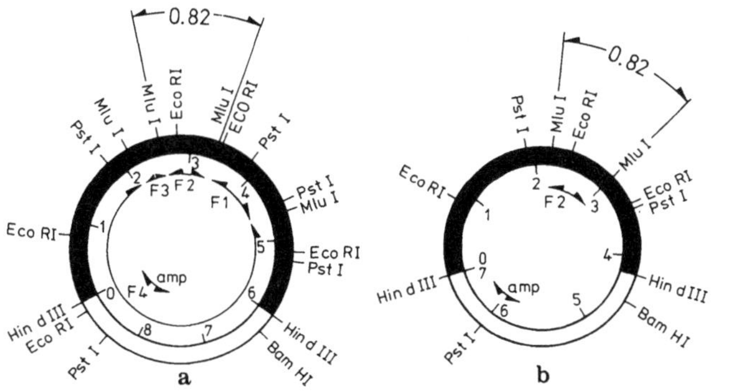

Fig. 2. Circular restriction maps of the hybrid plasmids pGK100 (A) and pGK100-5 (B) [8]. Hybrid plasmids were digested with various restriction endonucleases, and the molecular sizes of the linear DNA fragments generated were determined by agarose gel electrophoresis. The restriction sites were drawn to scale on a circular map. The number in the circle represents megadaltons. The 0.82-megadalton *Mlu*I fragment (F3) is depicted outside of the circle. (□) Vector plasmid pBR322; (■) chromosomal DNA fragment of *E. coli* B

products thus prepared were used for the transformation of *E. coli* 112L cells to select a hybrid plasmid carrying the *glk* gene. An increase in GLK activity was found only in the tranformants harboring the ligation product of F2 with F4. The introduction of the self-ligation product of F4 itself caused no increase in GLK activity. Therefore, the *glk* gene seemed to be located on a 0.82 MDa (F2) *Mlu*I fragment of the chromo-

Table 1. Effects of hybrid plasmids on Glk activity[a] and growth rate[b]

E. coli strain	Glk activity (μmole min^{-1} mg^{-1} of protein)	Generation time (h) on			
		Glucose	Glycerol	Fructose	Mannose
112L	0.012	33	2.4	2.7	15.8
112L(pGK100)	0.190	4.1	2.5	2.8	8.4
112(pGK100-5)	0.127	37	2.5	2.8	14.8
B	0.093	1.2	1.9	1.6	1.9
B(pGK100)	0.182	1.2	2.0	1.6	2.3
B(pGK100-5)	0.143	1.2	2.0	1.6	1.9

[a] Glk activities were assayed with extracts prepared from cells growing exponentially on Davis-Mingioli minimal medium (13) [0.7% KH_2PO_4, 0.3% K_2HPO_4, 0.1% $(NH_4)_2SO_4$, 0.01% $MgSO_4 \cdot 7\,H_2O$, 0.5% glucose] at 37 °C with shaking. Glk reaction was carried out in 3.0 ml of the mixture containing 50 mM glucose, 2.5 mM ATP, 2.0 mM $MgCl_2$, 0.25 mM NADP, 50 mM Tris-hydrochloride buffer (pH 7.5), 0.6 U of glucose-6-phosphate dehydrogenase, and cell extracts at 25 °C. Reaction was followed by recording the change in absorbance at 340 nm. The preparation of cell extracts was as described by Curtis and Epstein (12).

[b] The cells were grown on Davis-Mingioli minimal medium (13) containing various carbon sources at 0.5%. Incubation was carried out at 30 °C with shaking. Growth was monitored by measuring the turbidity at 610 nm

somal DNA segment in pGK100. The hybrid plasmid consisting of F2 and F4 was designated pGK100-5 (Fig. 2b).

Table 1 shows the GLK activities and growth rates of *E. coli* 112L and *E. coli* B with or without the hybrid plasmids [8]. In strains of 112L(pGK100) and 112L(pGK100-5), appreciably high GLK activities were observed (about 10- to 15-fold higher than that of 112L), although the effects of these hybrid plasmids on GLK activities were slight or only twice as high as in *E. coli* B cells. The poor growth rate of 112L on glucose was also improved by introducing pGK100. The generation time of 112L was 33 h on glucose at 30 °C, whereas that of strain 112L(pGK100) was only 4 h on glucose and about one-third of that of the wild *E. coli* B. Such an improvement of generation time was observed only when the cells were grown on glucose or mannose, there was no effect, however, when fructose or glycerol were employed as substrates. On the other hand, the hybrid plasmid pGK100-5 did not enhance the growth rate of 112L on glucose or mannose, although 112(pGK100-5) cells showed high GLK activity. These results indicated that the increase in GLK activity alone was not always sufficient to allow growth on glucose. Other factors coded by genes in the F1 or F3 region (Fig. 2) might be necessary for glucose catabolism by GLK.

To analyze products of the cloned gene, the cell extracts of 112L and 112L(pGK100) were subjected to DEAE-cellulose column chromatography and polyacrylamide gel electrophoresis [8]. The crude extract of 112(pGK100) showed a single large peak with GLK activity, whereas the activity in 112L was very low. Polyacrylamide gel electrophoresis of cell extracts of 112L(pGK100) showed a plasmid-dependent protein band having GLK activity when the gel slices were analyzed for enzymatic activity. Direct staining of the gel showed the existence of a single GLK band at the same position. Crude cell extracts of 112L(pGK100-5) also showed one plasmid-dependent protein band with GLK activity at the same position as that from 112L(pGK100).

Thus, we succeeded in cloning a gene responsible for GLK activity. Furthermore, factor(s) regulating the function of GLK in glucose catabolism seemed to be present in *E. coli*.

2.2 Phosphorylation of AMP to ATP by Dried Cells of Genetically Engineered *E. coli* Haboring the *glk* Gene

In microbial biocatalyst, the glycolytic pathway is a potent ATP regenerating process through which ADP is phosphorylated to ATP by the utilization of glucose and other energy sources. However, since dried *E. coli* cells cannot utilize glucose as described above, we have attempted to construct *E. coli* cells having a glucose phosphorylating activity by the introduction of the *glk* gene. It was hoped that genetically engineered *E. coli* cells harboring the *glk* gene would be more efficient at generating ATP from AMP.

2.2.1 Effect of Glucokinase Activity on ATP Production [14]

ATP production by dried *E. coli* cells was evaluated at optimum concentrations of glucose (0.1 M) and phosphate buffer (0.5 M). Fructose-1,6-bisphosphate (FBP) in the reaction mixture increased initially, reached a maximum, then declined. ATP was produced in association with the decrease in FBP concentration. To evaluate the effect of the cloned glucokinase activity on ATP production, the ATP-produc-

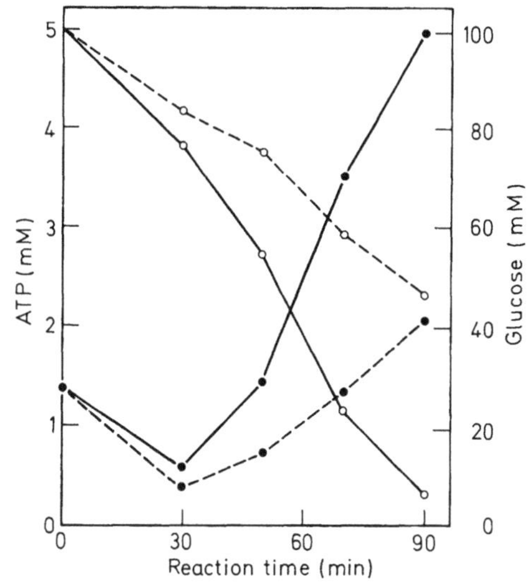

Fig. 3. Comparison of ATP-producing activity between *E. coli* B cells with and without pGK100. The reaction mixture contained 100 mM glucose, 30 mM AMP, 500 mM potassium phosphate buffer (pH 8.0), 5.0 mM $MgSO_4$, 1.0 mM ATP, 1.0 mM NAD and 50 mg ml^{-1} dried *E. coli* B cells with (———) or without (---------) hybrid plasmid pGK100. -●-, ATP; -○-, glucose

ing activity of dried control *E. coli* cells was compared with that of the cells carrying the hybrid plasmid (pGK100) [8,14], which harbors the glucokinase gene [8].

The introduction of the hybrid plasmid into *E. coli* cells resulted in an increase in glucokinase activity (1.5–1.7-fold) in both cell extracts and dried cells. The dried *E. coli* cells (50 mg ml^{-1}) with pGK100 showed about 2.5-fold higher ATP-producing activity than *E. coli* cells without pGK100 (Fig. 3). This result indicated that the glucokinase in dried cells was functioning in ATP production. Thus, the introduction of the cloned *glk* gene to *E. coli* cells seemed a promising approach for improving the ATP generation process in our microbiol biocatalyst system.

2.3 Cloning of the Phosphofructokinase (pfk) and Triosephosphateisomerase (tpi) Genes

Since phosphofructokinase (PFK) is a key enzyme in the glycolytic pathway and believed to be the primary site of allosteric regulation in the Pasteur effect, a cloned *pfk* gene was introduced and its effect on ATP production tested. Fraenkel et al. [15] first selected a hybrid ColEl plasmid (pLC16-4) containing both *pfk* and *tpi* (triose phosphate isomerase) genes from the Clarke and Carbon clone bank [16]. We determined the plasmid's effect [17] and restriction map [18], and used it for further experiments mentioned below.

2.3.1 Effect of Plasmid pLC16-4 on ATP Formation

The plasmid pLC16-4 was introduced into *E. coli* C600 by transformation. *E. coli* carrying this hybrid plasmid exhibited a 6-fold increase in PFK, or 11-fold increase in TPI which is another enzyme in the glycolytic pathway. The *pfk* and *tpi* genes are located very close to each other (between 87 to 88 min) in pLC16-4.

Dried cells prepared from the genetically engineered strain C600(pLC16-4) showed high ATP-forming activity [17]. Ninety five percent of the initial AMP (20 mM) was phosphorylated to ADP (20 %) and ATP (75 %) after 3 h of incubation, whereas with strain C600, only 50 % of the initial AMP was phosphorylated to ADP (25 %) and ATP (25 %). However, since pLC16-4 contained both the *pfk* and *tpi* genes closely linked to one another, it was important to elucidate the effect of the individual genes on ATP formation. To achieve this, the insert in pLC16-4 was subcloned in an effort to separate the *pfk* and *tpi* genes.

2.3.2 Subcloning of pLC16-4 [18] (Separation of *pfk* and *tpi* Genes)

Since the plasmid (pLC16-4) contained the structural genes for both *pfk* and *tpi*, attempts were made to separate the two genes. Thus, pLC16-4 was digested with five restriction endonucleases (*Eco*RI, *Hin*dIII, *Pst*I, *Bam*HI, and *Sal*I) and the fragments

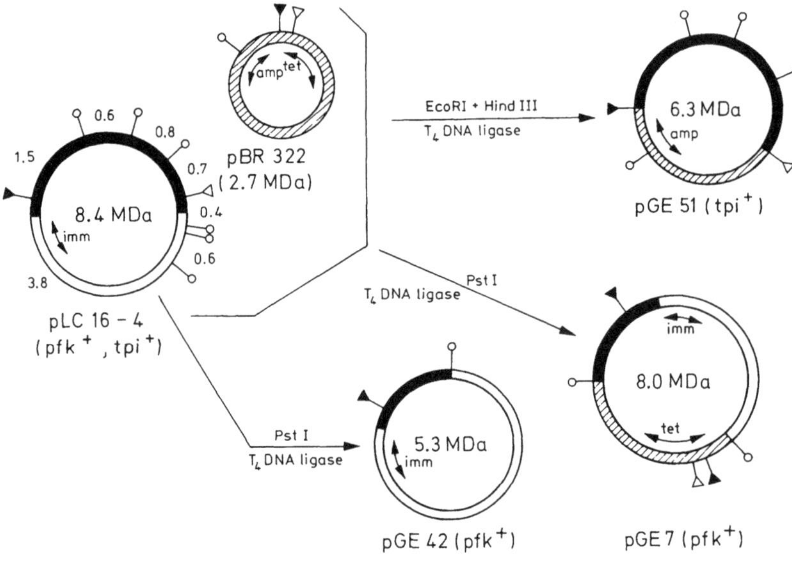

Fig. 4. Restriction cleavage map of pLC16-4 and schematic outline of recombinant plasmids pGE7, pGE42, and pGE51. Plasmid pLC16-4 has a molecular mass of 8.4 MDa and consists of *E. coli* chromosomal DNA (4.2 MDa) and ColE1 DNA (4.2 MDa). The numerals outside the circle show the molecular mass in MDa of the DNA fragments between each two restriction endonuclease-susceptible sites of pLC16-4. Each recombinant plasmid was constructed as follows: pGE7, the ligation between *Pst*I-digested pBR322 DNA and the *Pst*I-digested DNA fragment (5.3 MDa) containing the single *Eco*RI site of pLC16-4; pGE42, the self-ligation of the *Pst*I-digested 5.3-MDa fragment of pCL16-4; pGE51, the ligation between *Eco*RI-*Hin*dIII double-digested pBR322 DNA and the 3.6-MDa DNA fragment derived from *E. coli* chromosomal DNA of pLC16-4 with *Eco*RI-*Hin*dIII double digestion. Dark bands represent *E. coli* chromosomal DNA regions; white bands, ColE1 DNA regions; oblique line bands, pBR322 DNA regions. The molecular mass of each plasmid is indicated inside the circle. The marker gene of each plasmid is also indicated inside the circle: *amp* shows resistance to ampicillin; *tet*, resistance to tetracycline; *imm*, immunity to colicin E1. Symbols show restriction endonuclease-susceptible sites for *Eco*RI (▼), *Hin*dIII (▽), *Pst*I (○)

analyzed by agarose-gel electrophoresis. The plasmids was determined as molecular mass 8.4 MDa consisting of a DNA region (4.2 MDa) from the ColE1 plasmid and an *E. coli* chromosomal DNA region (4.2 MDa) inserted into the *Eco*RI site of the ColE1 DNA. The *E. coli* chromosomal DNA region consisted of two single sites for *Eco*RI and *Hin*dIII, respectively, and three sites for *Pst*I. No susceptible sites for *Bam*HI or *Sal*I were detected. Based on these data, the restriction map of pLC16-4 was constructed as shown in Fig. 4 [18].

To construct a recombinant plasmid harboring only one gene (either *pfk* or *tpi*), pLC16-4 was digested with restriction endonucleases, and fragments formed were cloned onto pBR322 (Fig. 4) [18]. The resulting hybrid recombinant plasmids were used to transform *E. coli* ET2036, which was deficient in both PFK and TPI activities. Deficiency in PFK and TPI activities resulted in a loss of the ability to grow on mannitol or glycerol, respectively. Transformants of *E. coli* ET2036 carrying either the *pfk* or *tpi* gene were selected on the basis of their ability to grow on either mannitol or glycerol, respectively. Growth of these transformants on the selective medium was inductive by the presence and expression of *pfk* and/or *tpi* genes.

A hybrid plasmid designated pGE51 was isolated which contained the 3.6 MDa DNA fragment derived from pLC16-4 by double digestion with *Eco*RI and *Hin*dIII (Fig. 4) [18]. The strain ET2036 carrying the pGE51 insert expressed only TPI activity, indicating that the 3.6 MDa DNA fragment contained only the *tpi* gene and that the *pfk* gene was cleaved by either *Eco*RI or *Hin*dIII. In order to preserve the *Eco*RI site, pLC16-4 was digested with *Pst*I and the longest *Pst*I-digested DNA fragment (5.3 MDa) was isolated and circulized (pGE42 in Fig. 4). Since this 5.3 MDa fragment contained on *ori* region of *Col*EI, it was capable of autonomous replication. This plasmid (pGE42) contained only the *pfk* gene. The *Pst*I-digested 5.3 MDa DNA fragment was also cloned into the *Pst*I site of pBR322 (pGE7 in Fig. 4). This plasmid (pGE7) also expressed only PFK activity. In spite of the fact that pGE7 contained two *ori* regions (one cloned from the pLC16-4 and the other from pBR322 DNA), it was stably maintained in *E. coli* strain C600.

Table 2. PFK and TPI activities of transformants[a]

Strain	Enzyme activity (μmol min^{-1} mg^{-1} of protein)	
	PFK	TPI
C600	0.45	3.32
C600(pLC16-4)	2.91	35.0
C600(pBR322)	0.47	4.33
C600(pGE7)	3.09	3.41
C600(pGE42)	2.39	2.33
C600(pGE51)	0.40	53.4

[a] Cells were grown in LB broth to early stationary phase.

2.3.3 Gene Dosage Effects of *pfk* and *tpi* on Enzyme Activity [18]

To estimate the gene dosage effects of hybrid plasmids pGE7, pGE42, and/or pGE51, each plasmid was introduced into *E. coli* C600, and the activities of PFK and TPI in the crude extracts of each transformant were assayed. Strains C600(pGE7) and C600(pGE42) showed 5- to 7-fold higher levels of activity for PFK than did strain C600 (control), but TPI activity remained unchanged as shown in Table 2. On the contrary, a C600(pGE51) strain showed a 16-fold higher level of TPI activity than did strain C600, although the level of PFK activity remained unchanged equally as in the control C600. These results clearly showed the existence of the *pfk* and/or *tpi* genes in each recombinant plasmid.

To ensure that the increase in PFK and TPI activities were due to gene dosage effects and not to the activation of the enzymes, crude protein extracts from strains C600 carrying each recombinant plasmid were analyzed by polyacrylamide gel electrophoresis. Strain C600(pLC16-4) showed two deep-strained protein bands corresponding to PFK and TPI activity, whereas C600(pGE7) and C600(pGE42) showed only a single protein band corresponding to PFK activity, and strain C600(pGE51) only one band for TPI activity [18].

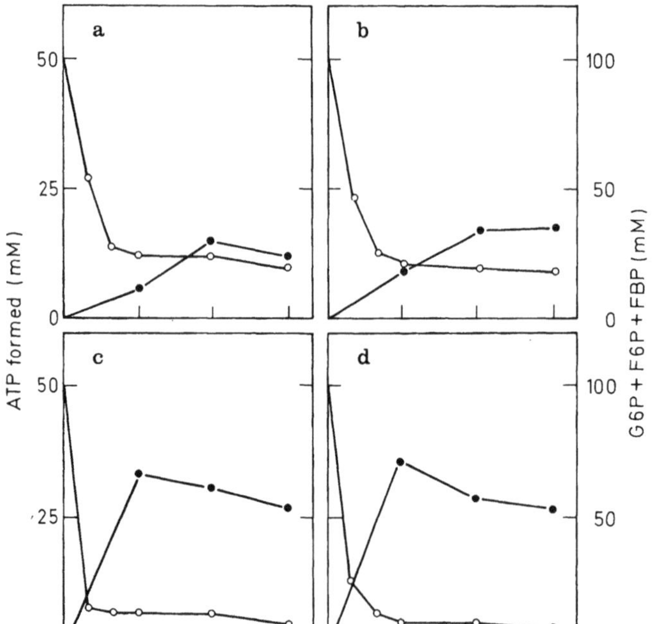

Fig. 5a–d. ATP formation with dried cells of *E. coli* C600 (**a**), C600(pBR322) (**b**), C600(pGE7) (**c**), and C600(pGE51) (**d**). At the prescribed times, the reaction was terminated, and the concentrations of ATP (●) and three hexosephosphates (G6P, F6P, FBP) (○) were determined

2.3.4 Effect of *pfk* and *tpi* on ATP Formation [18]

In the above Sections, construction of hybrid pBR322 plasmids carrying either the *pfk* or the *tpi* gene and the gene dosage effect resulting in expression of these genes has been described. Dried cells were prepared from C600(pGE7) and C600(pGE51) and analyzed for their ability to form ATP from AMP. The rate of AMP phosphorylation was accelerated in both cases (Fig. 5). In the reaction with dried cells of strains C600(pGE7) (Fig. 5c) and C600(pGE51) (Fig. 5d), about 70% of the initial AMP (50 mM) was phosphorylated to ATP after 1 h of incubation, while only 10% of the initial AMP was phosphorylated to ATP with dried cells of strain C600. Dried cells of strain C600(pBR322) (Fig. 5b) showed almost the same ATP formation activity as the control (C600) (Fig. 5a). The consumption of hexosephosphates (G6P, F6P, and FBP) in the reaction mixture of C600(pGE7) and/or C600(pGE51) were larger than those of the controls. In summary, construction of strains with supernumerary genes can effectively increase the efficiency of these cells to regenerate ATP.

3 Production of Glutathione by the "Syntechno System"

3.1 Production Strategy

Glutathione (GSH) is a detoxicating substance, and has many biochemical and pharmacological properties such as serving as a cofactor for many enzymes. The tripeptide, consisting of L-glutamate, L-cysteine, and glycine is biosynthetically made in two steps (Fig. 6). The first step is catalyzed by the enzyme GSH-I (γ-glutamylcysteine synthetase: E.C.6.3.2.2) and the second by GSH-II (glutathione synthetase: E.C.6.3.2.3). Since both enzymes require ATP, this biosynthetic pathway could be coupled with the ATP regenerating process described above.

This simple biosynthetic pathway is controlled at the cellular level by negative feedback inhibition. The first enzyme, GSH-I, is inhibited by glutathione (GSH) when large amounts of GSH accumulate in the cell. In order to use whole cells for GSH production, it was necessary to break this feedback inhibition cycle and select strains which could accumulate large amounts of GSH. Our strategy is shown in Fig. 6. Initially, mutant strains were isolated which maintained the catalytic activity of GSH-I but were desensitized for negative feedback inhibition. We isolated the gene (*gsh*-I) from the desensitized mutant and cloned it to pBR322 to make pBR322-*gsh*-I [19,20]. By the introduction of pBR322-*gsh*-I into *E. coli* C600 we could obtain colonies accumulating larger amounts of glutathione showing enhanced reactivity with nitroprusside-NH_3 on agar plates [20,21]. Having overcome the negative feedback inhibition of GSH-I by mutation and gene dosage, a second problem arose. The second step, catalyzed by GSH-II, became rate-limiting and an intermediate dipeptide γ-L-glutamyl-L-cysteine was accumulated in the reaction mixture [22]. In order to overcome this latter problem, we cloned the gene (*gsh*-II) coding for the second enzyme, GSH-II, into pBR322 to make pBR322-*gsh*-II [23]. Although introducing pBR322-*gsh*-II alone into the original *E. coli* C600 was ineffective for glutathione production, its co-introduction together with *gsh*-I was quite effective. From the genetically engineered

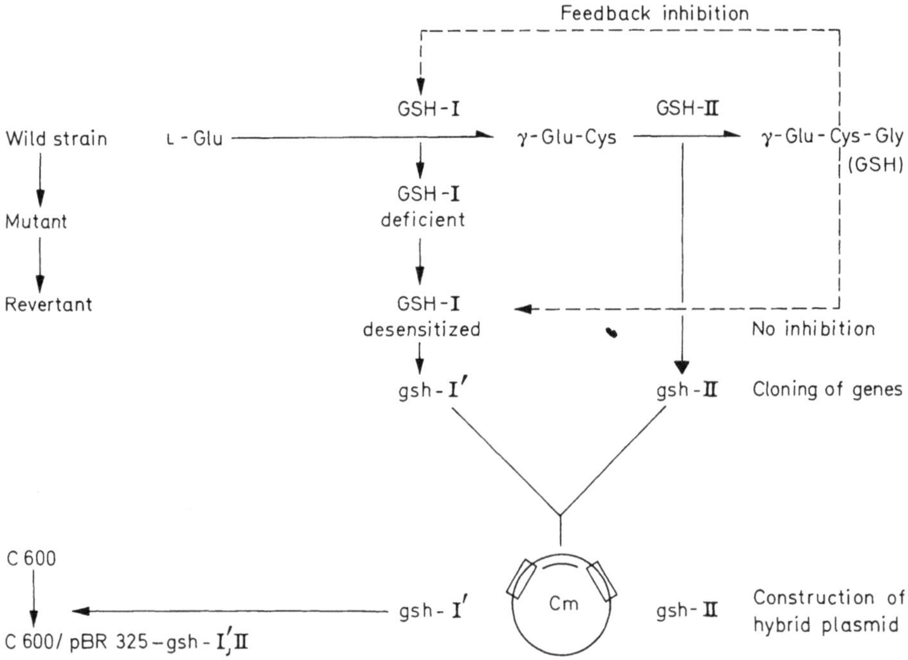

Fig. 6. Process strategy. In the wild strain (*E. coli* B), the first enzyme (GSH-I) is feedback inhibited by accumulated glutathione (GSH). An isolated mutant (C912) did not show any enzymatic activity (GSH-I), however, a revertant (RC912) obtained from C912 had a desensitized enzyme, which catalyzed the reaction but did not suffer from negative feedback inhibition. We isolated a gene (*gsh*-I') from the revertant. Symbol I' is used in this figure to indicate the desensitized gene, but actually the same gene with *gsh*-I. In the text, *gsh*-I' and *gsh*-I are used without distinction

E. coli cells, glutathione synthetase GSH-II was purified and characterized for the first time [24]. It had a molecular weight (MW) of 152,000 and was composed of four identical subunits each with a MW of 38,000.

Recently, the complete nucleotide sequence of the *gsh*-II gene has been determined and as a result, the primary structure of the GSH-II enzyme has been proposed [25]. It was confirmed by analyzing the sequence of 27 amino acids of the N-terminal of GSH-II.

In an attempt to further increase the efficiency of glutathione production, we constructed a hybrid plasmid containing both genes (pBR325-*gsh*-I,II) [26]. Introduction of this hybrid plasmid (pBR325-*gsh*-I,II) into *E. coli* was very effective for producing glutathione. The genetically engineered cells containing this hybrid plasmid always showed a constant ratio of both enzyme activities. One important consideration in the process is the efficiency of conversion of L-cysteine to glutathione, since L-Cys is the most expensive of the three component amino acids. The conversion efficiency increased and approached 100% when engineered cells which had been immobilized in carrageenan gel were used under suitable conditions.

When only one gene is considered, an increase in copy number may be sufficient to result in increased enzymatic activity. However, when two or more enzymes are involved, as in our case, the balance or ratio of the gene products becomes very important. We have succeeded in making various hybrid plasmids containing both genes (*gsh*-I,II) in various ratios [27], for example, (a) 2 *gsh*-I and 1 *gsh*-II (symbolized *gsh*-2I,II), (b) 1 *gsh*-I and 2 *gsh*-II (*gsh*-I,2II), (c) 1 *gsh*-I and 3 *gsh*-II (*gsh*-I,3II), etc. Using these genetically engineered cells, glutathione has been produced and it has been found that successful product formation is the result of both genetic engineering and process optimization (e.g., whole cell immobilization, process parameters, feed rates etc.). We propose, therefore, in the emerging field of biotechnology, processes which rely on a combined genetic engineering and process optimization to be termed "Syntechno Systems".

3.2 Cloning of a Gene (ghs-I) Responsible for GSH-I Activity

3.2.1 Glutathione-deficient Mutants

Two types of mutants being unable to synthesize glutathione were isolated [19]. Enzymatic analysis showed that one group (C912 and C915) lacked γ-glutamylcysteine synthetase (GSH-I) activity and the other (C1001) lacked glutathione synthetase (GSH-II) activity [19]. A revertant strain (RC912) of C912 had recovered full GSH-II activity and it was found to be completely free of negative feedback inhibition. Therefore, the gene (*gsh*-I) was cloned from this DNA extracted from RC912 [20].

3.2.2 Susceptibility of Mutants to Metal Compounds

Since mutants which are unable to synthesize glutathione can grow on minimal medium just like the wild-type strain, it was impossible to select transformants which were carrying the cloned *gsh*-I and/or *gsh*-II genes and were thus synthesizing glutathione [19]. In order to distinguish between them, we attempted to identify the glutathione-producing strains by their resistance to toxic substances. We concluded that there should

Table 3. *Minimum inhibitory concentrations of various compounds*

Compound	Minimum inhibitory concentration (μg ml^{-1})			
Strain	355	M910	C912	C1001
$HgCl_2$	4.0	5.0	0.1	0.4
Iodoacetamide	75	92	20	40
Tetramethylthiuram disulphide (TMTD)	90	100	5.5	10
8-Hydroxyquinoline (8-HQ)	15	20	4.0	10
Propyl 4-hydroxybenzoate	95	120	40	75
Methylglyoxal	25	40	5.5	7.5
Chloramphenicol	3.5	4.0	8.0	8.5
Dibromophenol	120	140	75	85
m-Cresol	1000	1000	900	1000
Tetrabromo-*m*-cresol	500	550	50	120
Thymol	250	300	150	200
Tetrabromophenolphthalein	95	130	55	75
Chloromethylketone	250	250	80	110

be some noticeable difference if glutathione was involved in detoxification. Chemicals which were tested included typical thiol-reactive agents ($HgCl_2$, iodoacetamide), a metal-chelating agent (8-hydroxyquinoline), an antibiotic (chloramphenicol), and others (mostly inhibitors of γ-glutamyltranspeptidase). As can be seen in Table 3, the GSH-I-deficient strain (C912) was more susceptible to various inhibitors with the exception of chloramphenicol. However, the GSH-II-deficient strain (C1001) was somewhat more resistant to these compounds. It would appear that the intermediate dipeptide L-Glu-L-Cys may be acting as a substitute for glutathione.

One striking difference which was observed between the mutants and the wild type involved cell viability. This was especially noticeable following exposure to tetramethylthiuram disulfide (TMTD). A marked decrease in viability of the GSH-I-deficient strain (C912) was observed. In agreement with the results of Apontoweil and Berends [21)], glutathione seemed to be a significant factor for the protection of cells exposed to toxic compounds. Based on the above observations, a selective medium was devised which contained sufficient amounts of TMTD to allow only the growth of the transformants.

3.2.3 Isolation of *gsh*-I with Various Restriction Endonucleases

Chromosomal DNA (1.2 μg) extracted from RC912 and plasmid vector pBR322 (1.0 μg) were digested, ligated, and used to transform strain C912. After 2 to 3 days, several TMTD-resistant colonies appeared. The number of colonies on the plate was dependent on both the kinds of restriction endonucleases used and the digestion time for chromosomal DNA (1.6 h). Transformants were detected by the formation of a red color in the presence of nitroprusside. Only 5–10% of TMTD-resistant colonies reacted with nitroprusside, and almost all colonies remained white-pink. The glutathione content and GSH-I activities of the transformed strains which did react with nitroprusside were analyzed. An increased accumulation of glutathione and an elevated GSH-I activity were observed in some strains. However, in other strains, both activities remained at the same level as those of the parental strain (M910). These tranformants might contain some genes directly involved with TMTD resistance.

3.2.4 Characterization and Analysis of *gsh*-I

Two kinds of hybrid plasmids carrying the *gsh*-I gene were obtained from transformants which made red colonies and designated as pGS100-1 and pGS100-2, respectively. Restriction analysis of pGS100-1 yielded the circular restriction map shown in Fig. 7 [22)]. pGS100-1 was found to be 4.7 MDa in size and contained a 2.1 MDa fragment of chromosomal DNA of strain RC912 cloned into the *Pst*I site of pBR322. pGS100-2 was 7.3 MDa in size and contained just the same fragment of chromosomal DNA as that found in pGS100-1. Restriction analysis of this hybrid plasmid indicated that pGS100-2 consisted of two molecules of pBR322 ligated at the *Pst*I site, indicated by P* in Fig. 7 (right). The chromosomal DNA fragment from RC912 was inserted between the second *Pst*I site generated by the ligation of the two molecules of pBR322 [22)]. The *Pst*I site designated by (P*) at the function of the two molecules of pBR322 is now resistant to *Pst*I attack, although the reason for this resistance is unknown. The whole DNA nucleotide sequence of *gsh*-I was only recently determined, and will be published soon.

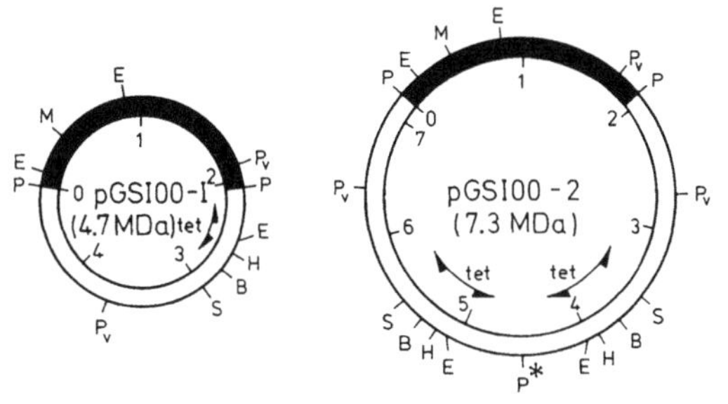

Fig. 7. Circular restriction maps of hybrid plasmid pBR322-*gsh*-I obtained from *E. coli* B. Circular restriction maps of pGS100-1 (Left) and and pGS100-2 (Right). The restriction sites are drawn relative to their scale on circular maps. The numbers in the circles represent megadaltons (MDa). P* in the map of pGS100-2 indicates the site resistant to *Pst*I attack. ■: *E. coli* B (RC912) chromosomal DNA; □: Vector plasmid pBR322. Abbreviated symbols: B, *Bam*HI; E, *Eco*RI; H, *Hin*dIII; M, *Mlu*I, P, *Pst*I; Pv, *Pvu*II; S, *Sal*I

3.2.5 Accumulation of the Dipeptide Intermediate γ-Glutamylcysteine by Cells Carrying the Cloned *gsh*-I Gene

E. coli B cells carrying the hybrid plasmid with the *gsh*-I gene had the ability to produce a large amount of glutathione from the constituent three amino acids (L-glutamate, L-cysteine, and glycine) in the presence of the ATP regeneration process consisting of the acetate kinase reaction in *E. coli*. In addition to glutathione, the intermediate dipeptide γ-glutamylcysteine (γ-GC) also accumulated [22]. It was of interest therefore, to investigate the production of γ-GC from L-glutamate and L-cysteine by the genetically engineered (C1001/pGS100-1) cells which were deficient in *gsh*-II. γ-GC was efficiently produced by the cells of strain C1001/pGS100-1, and 20 mM L-cysteine was completely converted to γ-GC in the presence of an excess of L-glutamate (80 mM). This unusual requirement for high levels of L-glutamate was presumably due to the utilization of L-glutamate for other than glutathione production.

3.3 Cloning of the gsh-II Gene for GSH-II Activity

When the *gsh*-I gene for GSH-I was cloned and introduced into *E. coli* B cells, GSH-I activity increased. However, the second reaction catalyzed by GSH-II appeared to become the rate-limiting step in glutathione production and consequently, the intermediate dipeptide γ-GC accumulated. In order to overcome this problem in glutathione production, the *gsh*-II gene [23] was cloned into *E. coli*.

3.3.1 Isolation of *gsh*-II in an Effort to Construct a More Efficient Glutathione Producer

Chromosomal DNA was extracted from *E. coli* strain RC912, a mutant synthesizing GSH-II which is resistant to feedback inhibition. The DNA was digested with *Hin*dIII restriction endonuclease and ligated into pBR322 which had been previously treated with *Hin*dIII in combination with bacterial alkaline phosphatase according to the method of Ullrich et al. [28]. To select for the hybrid plasmid harboring the *gsh*-II gene, cells of strain C1001 (an *E. coli* B mutant deficient in GSH-II activity) were transformed with the complete ligation mixture by the method of Kushner [29]. Transformants were spread on Davis-Mingioli minimal medium supplemented with 4.0 μg ml^{-1} ampicillin and 80 μg ml^{-1} tetramethylthiuram disulfide (TMTD) or a growth inhibitor of strains deficient in glutathione biosynthesis. By this method, 21 TMTD-resistant colonies presumably harboring the *gsh*-II-containing plasmid, were obtained [23] and the frequency of the appearance of TMTD-resistant colonies was about 2×10^{-7}. To ascertain whether these colonies contained the hybrid plasmids harboring *gsh*-II, the amount of glutathione produced by the transformants was determined. Only three colonies showed appreciably higher amounts of glutathione than the control C600 although all colonies contained detectable levels of this tripeptide. These results suggest that in addition to the three transformants which presumably carry the hybrid plasmid encoding the GSH-II activity, other genes may be present which encode a TMTD resistance not linked to GSH-II production. In order to analyze the extrachromosomal DNA, cleared lysates were prepared from each of the transformants.

3.3.2 Characterization and Analysis of *gsh*-II

Two kinds of hybrid plasmids were isolated from transformants which showed high GSH-II activity and were characterized with various restriction endonucleases. The smaller one obtained with *Hin*dIII was 4.2 MDa in molecular weight, involving the 1.6 MDa chromosomal DNA fragment of *E. coli* in the *Hin*dIII site of pBR322. On the

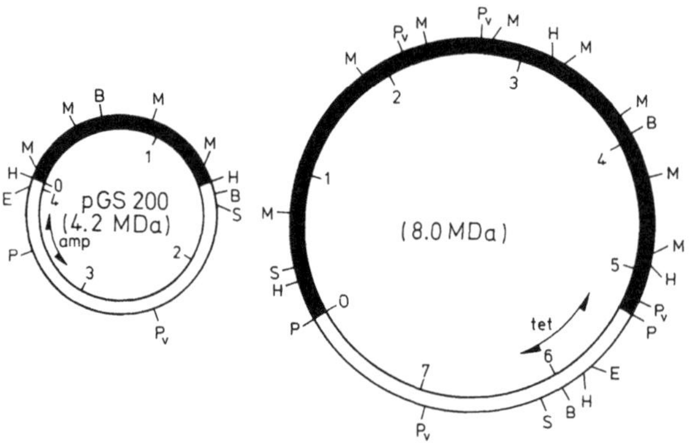

Fig. 8. Circular restriction maps of hybrid plasmid pBR322-*gsh*-II obtained from *E. coli*

other hand, the larger one obtained with *Pst*I was 8.0 MDa, containing the 4.5 MDa fragment of chromosomal DNA in the *Pst*I site of pBR322. Based on the restriction enzyme digestion patterns, the circular maps of these hybrid plasmids were constructed (Fig. 8). Although the two hybrid plasmids were obtained by cloning into two different restriction sites, both contained the complete *gsh*-II gene. The smaller *Hin*dIII fragment (1.6 MDa) was found to coincide with a portion of the larger cloned fragment (4.5 MDa). This smaller fragment was designated pGS200 and was used to subclone the entire *gsh*-II gene [30] which recently has been completely sequenced [25].

3.3.3 Subcloning and Stabilization of *gsh*-II

To obtain the *gsh*-II gene itself, we attempted to subclone the fragment containing the *gsh*-II gene from the 4.2 MDa hybrid plasmid (pGS200) into the *Hin*dIII and *Bam*H3 sites of pBR325. The gene appeared to be located counterclockwise from the

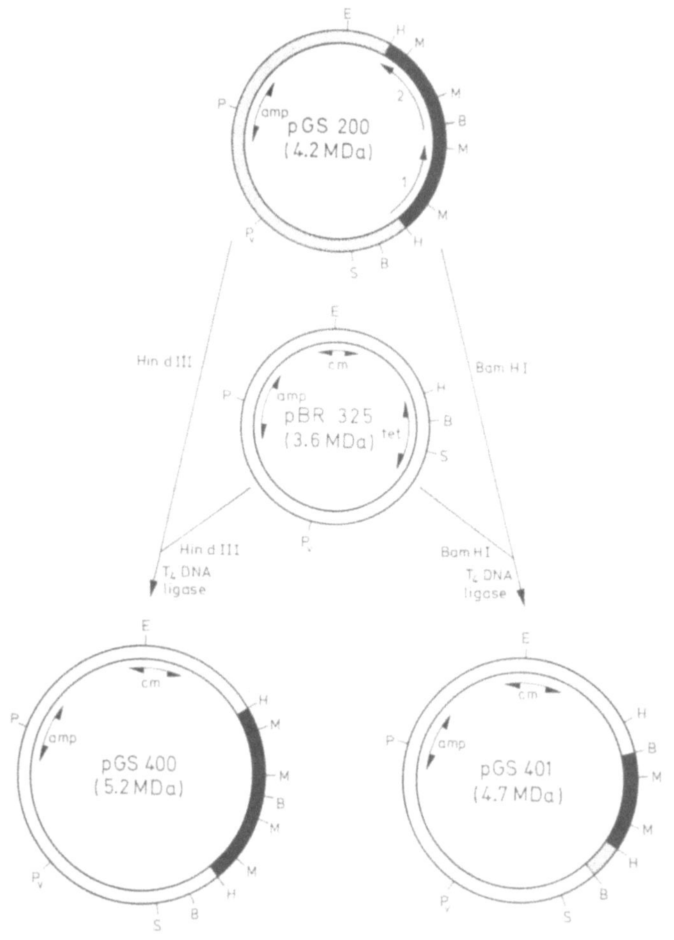

Fig. 9. Stabilization of *gsh*-II. The plasmid pGS200 containg *gsh*-II was very unstable. It contained an unidentified gene (indicated by the arrow 2 in pGS200) in addition to *gsh*-II (shown by arrow 1). When *gsh*-II was transferred to pBR325, the resulting pGS401 became very stable, while pGS400 remained unstable. Therefore, we believe a certain base sequence, indicated by arrow 2, is responsible for the instability of the plasmid

*Hin*dIII site to the *Bam*HI site (Fig. 9) in the original plasmid. This hybrid plasmid (pGS200) was originally very unstable. However, removal of the superfluous DNA between the *Bam*HI and *Hin*dIII sites (denoted by arrow 2 in pGS200, Fig. 9) resulted in a stable plasmid [30]. The subcloned DNA fragment of 1.1 Md in pGS401 (Fig. 9) seemed almost to correspond to the required amount of DNA to code for the GSH-II enzyme, whose molecular weight was determined to be 152,000. This DNA fragment was subjected to base sequence analysis by the "Dideoxy sequencing method" of Sanger et al. [31]. Further investigation revealed that the DNA fragment removed from pGS200 during stabilization (fraction between *Bam*HI and *Hin*dIII in Fig. 9) encoded a peptide [32], which when expressed confers an unstable phenotype on the cell. Further characterization of this peptide is currently under way.

3.3.4 DNA Base Sequence of *gsh*-II

The hybrid plasmid containing the subcloned *gsh*-II gene was designated pGS401 in Fig. 9 and subjected to structure analysis or base sequencing. The complete nucleotide sequence revealed that it had typical promotor and termination signals [25]. The enzyme (GSH-II) coded by *gsh*-II has been purified about 60-fold [24] and subjected to various analyses. The first 27 amino acids (excluding amino acid 21) were determined by amino acid sequence analysis and found to coincide completely with the nucleotide sequence. Amino acid 21 was deduced to be serine from the DNA sequence data.

3.4 Construction of a Hybrid Plasmid Containing both *gsh*-I and II in Various Ratios

At the beginning of this work, the *gsh*-I and II genes were separately cloned into pBR322 and introduced into *E. coli* cells stepwisely or at the same time. However, the ratio of both enzyme activities varied from experiment to experiment. Therefore, to keep their ratio constant we tried to construct a hybrid plasmid harboring both genes in the same plasmid in a ratio of 1 : 1 [26], then later in various ratios [27, 33].

3.4.1 Construction of a Hybrid Plasmid pGS500 Containing *gsh*-I and *gsh*-II

Figure 10 shows a scheme for the construction of the hybrid plasmid pGS500 having the two genes *gsh*-I *and* II for both enzymes on the same plasmid vector [26]. In this case the vector was pBR325 which contains the chloramphenicol resistance marker in addition to the tet and amp markers. For the construction of pGS500, two kinds of hybrid plasmids, pGS100-2 and pGS200, were used. As shown in Fig. 10, pGS100-2 (7.3 MDa) consisted of a dimer of pBR322 with the insert of chromosomal DNA fragment (2.1 MDa) harboring the *gsh*-I in the *Pst*I site generated during dimer formation. Fortunately, the *Pst*I site shown by P* became inert and was not cleaved during the treatment with *Pst*I, so the fragment containing *gsh*-I was easily separated from the dimer of pBR322 by agarose gel electrophoresis. The pGS200 was 4.2 MDa in size and contained an *E. coli* chromosomal DNA fragment (1.6 MDa) encoding the *gsh*-II gene in the *Hin*dIII site of pBR322.

The hybrid plasmid pGS100-2 was digested with *Pst*I and the DNA fragments (2.1 MDa) containing the *gsh*-I gene were obtained by extraction from agarose gel after electrophoresis of the digestion mixture. The linear fragments obtained were annealed with pBR325 pretreated with *Pst*I and ligated with T_4 DNA ligase. This ligation mixture was used to transform *E. coli* strain C912 (*gsh*-I^-) to C912 (*gsh*-I^+). These transformants were selected as colonies resistant to both TMTD and tetracycline. The hybrid plasmid contained in one of these transformants was isolated and designated pGS300. In a similar manner, the *gsh*-II gene was introduced into pGS300. pGS200 was digested with *Hin*dIII and the *E. coli* DNA fragment (1.6 MDa) containing the *gsh*-II was obtained by electrophoresis. This linear DNA fragment containing the *gsh*-II gene was annealed with pGS300 pretreated with *Hin*dIII and ligated with T_4 DNA ligase. The plasmid thus constructed could transform C1001 (*gsh*-II^-) to C1001 (*gsh*-II^+). Transformants having a hybrid plasmid containing both *gsh*-I and *gsh*-II

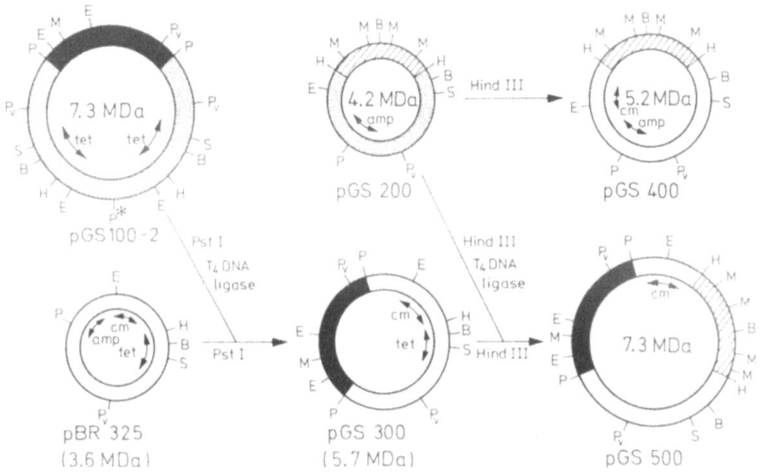

Fig. 10. Scheme for construction of hybrid plasmids pGS300, pGS400, and pGS500. For the subcloning of the genes (*gsh I* for GSH-1, *gsh II* for GSH-II), two kinds of hybrid plasmids, pGS100-2 and pGS200, were used. The structural and functional properties of these hybrid plasmids were shown in our previous papers (20, 22, 23). P* in pGS100-2 shows the site resistant to *Pst*I attack. The DNA fragment with *gsh* I was isolated from the *Pst*I digestion mixture of pGS100-2. pGS300 was constructed by inserting this DNA fragment into the *Pst*I site of pBR325. Similarly, the DNA fragment with the *gsh II* gene was isolated from the *Hin*dIII digestion mixture of pGS200. pGS400 and pGS500 were constructed by inserting this DNA fragment into the *Hin*dIII sites of pBR325 and pGS300, respectively. For other detailed conditions see Results. ■, *E. coli* B (RC912) chromosomal DNA fragment with *gsh* I gene; ▥, *E. coli* B (RC912) chromosomal DNA fragment with *gsh* II gene; ▩, vector plasmid pBR322; □, vector plasmid pBR325. Symbols: P, *Pst*I; E, *Eco*RI; B, *Bam*HI; S, *Sal*I; M, *Mlu*I; H, *Hin*dIII; Pv, *Pvu*II. *Amp*, *tet*, and *cm* show the genes for the resistance to ampicillin, tetracycline, and chloramphenicol, respectively

genes were selected as colonies resistant to both TMTD and chloramphenicol. The hybrid plasmid isolated from one of these transformants was designated pGS500. The linear DNA fragment prepared from pGS200 was also annealed with pBR325 pretreated with *Hin*dIII, ligated with T_4 DNA ligase, and used for the transformation of strain C1001 (*gsh*-II$^-$) to C1001 (*gsh*-II$^+$). The transformants harboring the hybrid plasmid containing only the *gsh*-II gene were selected as the colonies resistant to both TMTD and ampicillin. The hybrid plasmid isolated from one of these transformants was designated pGS400.

To ascertain the existence of *gsh*-I and *gsh*-II on these hybrid plasmids (pGS300, pGS400, and pGS500) and to evaluate the effects of these hybrid plasmids on GSH-I and GSH-II activities, the hybrid plasmids were introduced into cells of strain C600 (restriction$^-$ and modification$^-$) (Table 4). The introduction of pGS300 and pGS400 resulted in a marked increase in GSH-I (11-fold) and GSH-II (14-fold) activities, respectively, in comparison with the activities of C600 alone. The co-introduction of pGS300 and pGS400 was also effective in increasing GSH-I and GSH-II activities, but the ratio of both activities was not always constant. In contrast to the co-introduction of both hybrid plasmids, the introduction of the hybrid plasmid pGS500 carrying both genes on the same vector resulted in a constant increase in the ratio of both enzyme activities.

Table 4. Enzymatic activities of prototrophic and recombinant strains

Strain C600/	Hybrid plasmid	Enzyme activity[a]		GSH produced[b]
		GSH-I	GSH-II	
C600	—	0.057 (1.0)	0.641 (1.0)	8.1 (1.0)
C600/pGS100	pBR322-*gsh*-I	1.778 (31.2)	0.519 (0.8)	—
C600/pGS300	pBR325-*gsh*-I	1.370 (24.0)	0.665 (1.0)	—
C600/pGS400	pBR325-*gsh*-II	0.052 (0.9)	16.30 (25.4)	—
C600/pGS401	pBR325-*gsh*-II′	0.067 (1.2)	15.57 (24.3)	—
C600/pGS500	pBR325-*gsh*-I,II	1.723 (30.2)	16.05 (25.0)	255.4 (31.5)
C600/pGS330	pBR325-*gsh*-I,I	3.071 (53.9)	1.02 (1.6)	26.1 (3.2)
C600/pGS501	pBR325-*gsh*-I,II′	2.258 (39.6)	15.29 (23.8)	234.5 (29.0)
C600/pGS550	pBR325-*gsh*-I,I,II	2.857 (50.1)	10.07 (15.7)	390.9 (48.3)
C600/pGS551	pBR325-*gsh*-I,I,II′	3.913 (68.6)	16.33 (25.5)	471.7 (58.6)

[a] protein (mole mg^{-1} h^{-1})
[b] wet cell (mole g^{-1} h^{-1})
II′: pGS401, II: pGS200

3.4.2 Construction of a Hybrid Plasmid Containing *gsh*-I and -II in Various Ratios

We have also constructed hybrid plasmids containing both the *gsh*-I and -II genes in various ratios, for example, 1:2, 1:3, 2:1, etc. [27)]. Increasing the number of copies of the *gsh*-II gene (for example, 1:2 and 1:3) showed an increase in GSH-II activity but did not result in increased glutathione production. This result can be traced to the fact that in these transformants GSH-I activity was rate limiting. Introduction of a hybrid plasmid containing *gsh*-I and *gsh*-II in a ratio of 2:1 produced the highest amount of glutathione as shown in Table 4 [33)].

In these cases, the problem was that these hybrid plasmids containing many redundant genes were rather unstable, and converted to the smaller-size plasmid.

In order to make genes expressed more effectively, we are manipulating the promoter region. The result will be published elsewhere in the near future.

3.5 Production of Glutathione by the "Syntechno System" with Genetically Engineered Cells

Glutathione production by *E. coli* B cells with hybrid plasmids was carried out in a bioreactor or biocatalyst system with an ATP regeneration reaction catalyzed by acetate kinase in *E. coli* B cells. The introduction of pGS300 (containing *gsh*-I) into *E. coli* cells remarkably increased GSH-I activity and, at the same time, glutathione-producing activity. However, cells containing only pGS400 (*gsh*-II) did not show enhanced glutathione production, although they did exhibit increased *gsh*-II activity. The introduction of pGS400 together with pGS300 could effectively increase glutathione production. However, the ratio of the two activities varied resulting in inconsistant yield of the product. Introduction of pGS500, which possessed both genes (*gsh*-I and II) on the same plasmid, into *E. coli* always yielded a constant ratio of both enzyme activities, and cells containing the pGS500 produced a large amount of glutathione in the reaction mixture.

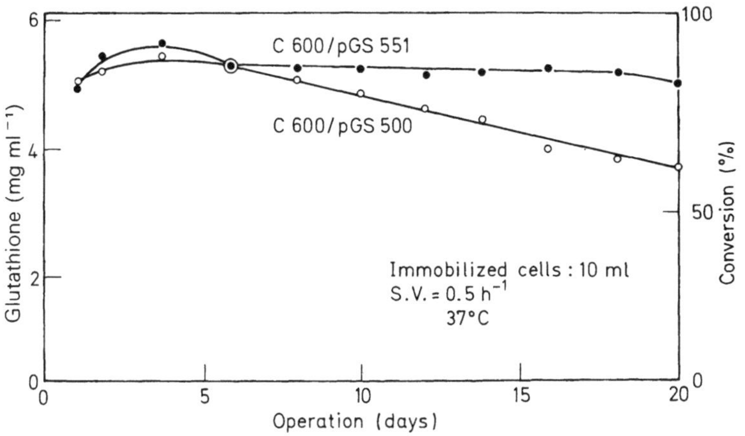

Fig. 11. Continuous production of glutathione by immobilized *E. coli* cell column. Substrate solution: 80 mM L-glutamate, 20 mM L-cysteine, 20 mM glycine, 25 mM $MgCl_2$, 25 mM K-phosphate buffer (pH 7.5), 1.0 mM ATP and 59 mM acetyl phosphate

In the production of glutathione, the conversion efficiency (%) of L-cysteine to glutathione is very important, in that L-cysteine is the most expensive amino acid of the three components. The conversion efficiency increased and reached 100% when the recombinant microorganisms were immobilized in carrageenan gel and used under optimal conditions. We attempted to further optimize the conditions by carrying out predetermined feeding regimes or continuous culture, etc. In Fig. 11, an example of continuous culture is shown. In this case, more than 5 g L^{-1} of glutathione (conversion efficiency from L-cysteine was about 85%) were produced by C600/pGS551 and continuous production could be maintained for 25 days. The pGS551

Table 5. Contribution of various techniques to GSH production

Strain	GSH ($g\,l^{-1}$)	Conversion from L-Cys	Notes
C600	0.2	10%	
C600-*gsh*-I	1.2	20	[a,b]
C600-*gsh*-II	0.3	10	[a,b]
C600-*gsh*-I,II	2.2	35	[b]
C600-*gsh*-I,II	3.3	52	[b]
C600-*gsh*-I,II	4.5	75	L-Glu 80 mM[c]
C600-*gsh*-I,II	6.0	100	Imm. Cys 20 mM[c,d]
C600-*gsh*-I,II	11.3	94	Imm. Cys 40[c,d]
C600-*gsh*-I,II	20.3	85	Imm. Cys 80[c,d]
(C600	0.6	10	Imm. Cys)

[a] Fundamental knowledge of microorganisms
[b] Genetic engineering
[c] Bioprocess
[d] Immobilization

contains two *gsh*-I and one stabilized *gsh*-II. Efforts are currently under way to commercialize this process.

In Table 5, the increase in the yield of glutathione and the conversion efficiency from L-cysteine to glutathione under various conditions are summarized. From Table 5 it appears that genetic engineering techniques are not always necessary. The yield of glutathione increased from 0.2 to 3.3 g L^{-1} by the gene engineering technique, yet it was elevated to 20 g L^{-1} by other techniques such as submerged culture and immobilization, etc. However, when we carried out the same experiments using original prototroph strains under optimal conditions, we were unable to obtain a high glutathione yield, (i.e., 0.6 g L^{-1} as shown at the bottom or in parentheses in Table 5). This figure represents only a 3-fold increase over C600 in batch culture. Therefore, the combination of all these techniques (recombinant DNA, immobilization, submerged culture) and fundamental knowledge in microbiology (for example, on negative feedback inhibition) are very important, in achieving high yield. We propose to call such a combination a "Syntechno System".

4 Future Aspects (Protein Engineering)

One of the future problems is to connect the glutathione production effectively with the ATP regeneration process mentioned in Fig. 1. Recently, Fujio et al. [34] developed an intact cell system, where ATP was supplied indirectly by adding glucose, in which glutathione was effectively produced. This system will be economically effective, and therefore, we are now trying to construct a hybrid plasmid carrying some genes of the glycolytic pathway together with genes (*gsh*-I,II) for glutathione synthesis. As mentioned above, we have cloned some genes for various enzymes, and are trying to determine their DNA base sequences. Some DNA base sequences, for example, that of *gsh*-II, has been already determined, and based on this the amino acid sequence of the enzyme GSH-II was deduced. Base and amino acid sequences were also determined quite recently. These results are being processed by a computer, and the enzyme structure is now under investigation. Using site-directed mutagenesis, we are planning to exchange some bases of the DNA, and, as a result, to exchange the specific amino acids in the enzyme. This kind of research will be quite useful to change the specificity of enzyme and to make it more effective [35]. By the combination of knowledge in different fields it is believed that protein engineering could be feasible in the near future.

5 References

1. Kimura, A.: Proceeding of the IVth Internat. Symp. on "Genetics of Industrial Microorganisms" p. 277. Kyoto: Kodansha 1982
2. Kimura, A., Hirose, K., Kariya, Y., Nagai, S.: J. Bacteriol. *125*, 744 (1976)
3. Umemura, I., Fukuda, H., Kimura, A.: Eur. J. Appl. Microbiol. Biotechnol. *15*, 133 (1982)
4. Kimura, A., Okuda, M.: Agric Biol. Chem. *40*, 1373 (1976)
5. Kimura, A., Tatsutomi, Y., Fukuda, H., Morioka, H.: Biochim. Biophys. Acta *629*, 217 (1980)
6. Kimura, A., Okuda, M., Fukuda, H.: J. Appl. Biochem. *1*, 127 (1979)
7. Asada, M., Shirai, Y., Nakanishi, K., Matsuno, R., Kimura, A., Kamikubo, T.: J. Ferment. Technol. *59*, 239 (1981)

8. Fukuda, Y., Yamaguchi, S., Shimosaka, M., Murata, K., Kimura, A.: J. Bacteriol. *156*, 922 (1983)
9. Fukuda, Y., Yamaguchi, S., Shimosaka, M., Murata, K., Kimura, A.: Agric. Biol. Chem. *48*, 2541 (1984)
10. Fukuda, Y., Yamaguchi, S., Hashimoto, H., Shimosaka, M., Kimura, A.: ibid. *48*, 2877 (1984)
11. Postma, P. W., Roseman, L. S.: Biochim. Biophys. Acta *457*, 213 (1976)
12. Curtis, S. J., Epstein, W.: J. Bacteriol. *122*, 1189 (1975)
13. Davis, B. D., Mingioli, E. S.: ibid. *60*, 17-28 (1950)
14. Yamaguchi, S., Fukuda, Y., Shimosaka, M., Kimura, A.: J. Ferment. Technol. *62*, 29 (1984)
15. Thomson, J., Gerstenberger, P. D., Goldberg, D. E., Gociar, E., Silva, A. O., Fraenkel, D. G.: J. Bacteriol. *137*, 502 (1979)
16. Clarke, L., Carbon, J.: Cell *9*, 91 (1976)
17. Shimosaka, M., Fukuda, Y., Kimura, A.: Agric. Biol. Chem. *45*, 1025 (1981)
18. Shimosaka, M., Fukuda, Y., Murata, K., Kimura, A.: J. Bacteriol. *152*, 98 (1982)
19. Murata, K., Kimura, A.: J. Gen. Microbiol. *128*, 1047 (1982)
20. Murata, K., Kimura, A.: Appl. Environ. Microbiol. *44*, 1444 (1982)
21. Apontoweil, P., Berends, W.: Biochem. Biophys. Acta *399*, 10 (1975)
22. Gushima, H., Miya, T., Murata, K., Kimura, A.: Agric. Biol. Chem. *47*, 1927 (1983)
23. Murata, K., Miya, T., Gushima, H., Kimura, A.: ibid. *47*, 1381 (1983)
24. Gushima, H., Miya, T., Murata, K., Kimura, A.: J. Appl. Biochem. *5*, 210 (1983)
25. Gushima, H., Yasuda, S., Soeda, E., Kondo, M., Kimura, A.: Nucl. Acid. Res. *12*, 9299 (1984)
26. Gushima, H., Miya, T., Murata, K., Kimura, A.: J. Appl. Biochem. *5*, 43 (1983)
27. Watanabe, K., Yamano, Y., Kimura, A.: Abstract of Fermentation Technology Symp., p. 32. 1984 Osaka
28. Ullrich, A., Shine, J., Chirgwin, J., Pictet, P., Tischer, E., Rutter, W. J., Goodman, H. M.: Science *196*, 1313 (1977)
29. Kushner, S. R.: Genetic Engineering (eds. H. W. Boyer, S. Nicosia) pp. 17–128. Elsevier North Holland, Biochemical Press 1981
30. Miya, T., Gushima, H., Kimura, A.: (unpublished data)
31. Sanger, F., Nicklen, S., Coulson, A. R.: Proc. Natl. Acad. Sci. USA. *74*, 5463 (1977)
32. Watanabe, K., Saito, H., Yamano, Y., Kimura, A.: (unpublished data)
33. Miya, T., Kimura, A.: (unpublished data)
34. Fujio, T., Hayashi, M., Tomiyoshi, Y., Fujioka, A.: Japanese Patent, shouwa 60-105499 (1985)
35. Craik, S. C., Largman, C., Fletcher, T., Roczniak, S., Barr, P. J., Fletterick, R., Rutter, W. J.: Science *228*, 291 (1985)

Extracellular Microbial Lipids as Biosurfactants

Dieter Haferburg, Rolf Hommel[1], Reiner Claus and Hans-Peter Kleber
Bereich Biochemie, Sektion Biowissenschaften, Karl-Marx-Universität Leipzig, 7010 Leipzig, GDR

1 Introduction ... 54
2 Synthesis and Composition of Biosurfactants ... 55
2.1 Microorganisms and Conditions for Synthesis ... 55
2.2 Structure and Properties ... 67
2.2.1 Structure ... 67
2.2.2 Properties ... 71
3 Biosynthesis of Biosurfactants ... 71
3.1 Acylated Sugar Derivatives ... 71
3.1.1 Glycolipid Derivatives ... 74
3.1.2 Mycolic Acid Derivatives ... 75
3.2 Other Biosurfactants ... 76
4 Physiological Importance of Biosurfactants ... 76
4.1 Biosurfactants and Hydrocarbon Assimilation ... 77
4.1.1 Theoretical Considerations and Practical Relevance ... 77
4.1.2 Effect of Surface-Active Agents on Hydrocarbon Assimilation ... 78
4.2 Biological Activities of Biosurfactants ... 81
4.2.1 Mycolic Acid-Type and Similar Biosurfactants from *Actinomycetes* ... 81
4.2.2 Biocidic Effects of Biosurfactants from Non-*Actinomycetes* ... 82
5 Applications and Concluding Remarks ... 85
6 Nomenclature and Definitions ... 88
7 References ... 88

During the past 10 years there have been rapid developments in the detection and characterization of surface-active agents formed by bacteria, yeasts and fungi. The present paper reviews the available information on microbial processes which result in the formation of such compounds — biosurfactants —, i.e. appropriate conditions as well as the known chemical structures and physico-chemical properties of isolated biosurfactants. The physiological importance of extracellular surface-active compounds is discussed, particularly with respect to their possible role in the microbial degradation of water-insoluble carbon-sources and their possible biocidic effects. A survey on commercial applications is given and conclusions for further research work are presented.

[1] Institut für Technische Mikrobiologie, Außenstelle Leipzig, 7010 Leipzig, PSF 674, GDR

1 Introduction

Surface-active compounds (detergents, surfactants, amphiphiles) are molecules which consist of both hydrophilic and hydrophobic moieties. They possess the capability to reduce the surface tension (air-water) of aqueous media, to reduce the interfacial tension of liquid-liquid (e.g. oil-water) or liquid-solid (e.g. wetting phenomena) systems, and to effect changes in the foaming properties of aqueous mixtures [1]. By reason of these properties, surface-active compounds are able to facilitate reactions and mass transfer at the interphase. The economic importance of synthetic surfactants is reflected by the increasing number of publications and patent claims as well as by increasing consumption on a world-wide scale. In 1980 in North America alone, the surfactant market amounted to more than 2.5×10^6 metric tons [2]. Some of these synthetic surfactants are toxic and not readily biodegradable. No single surfactant is suitable for all required applications. Therefore, alternative sources of surfactants or surfactant precursors are a necessity.

During recent years a considerable number of studies on surfactants produced by microorganisms have been published. The advantages of these biosurfactants are that new chemical types can be introduced, which would be difficult to synthesize, and that these compounds can be produced relatively simply and inexpensively. Furthermore, most of these microbial biosurfactants are biodegradable.

Most biosurfactants are exolipids corresponding from a physicochemical point of view to their chemical counterparts. Their biosynthesis by microorganisms is a particular biochemical/microbiological field as well as being a new biotechnological tool. In many cases extracellular lipids are not components of membranes; the term "Lipid" defines only the amphiphilic character of the compounds. Biosurfactants and their enzyme-catalyzed synthesis are related to the chemical structure and the sequence of reactions fixed by the genetic code: Any change of the amphiphilic constitution of the molecule is only possible within a small range of the corresponding homologous compounds. The excretion can be detected in certain phases of the life-cycle (late exponential phase/stationary phase) and probably depends on the metabolic state of the cell which is also reflected by the morphology of the cell-boundary. Moreover, excretion is strongly influenced by suitable methods of cultivation.

The role of exolipids during the microbial utilization of n-alkanes was a direct starting point for the biotechnological process: Enhanced oil recovery seems to be the

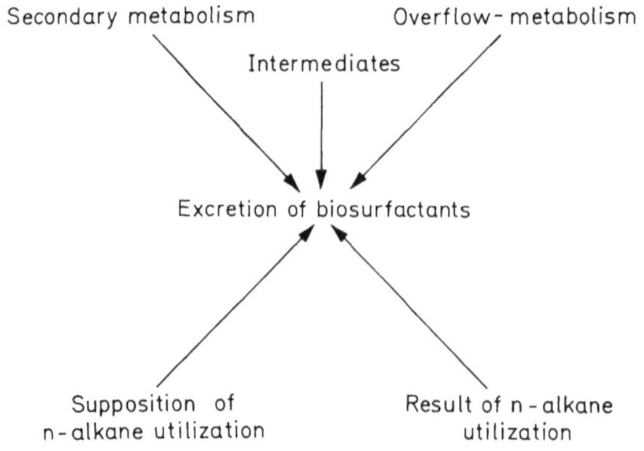

Fig. 1. Possible routes of biosurfactants generation

favoured field of application. The terminology of these microbial extracellular metabolites — biosurfactants, bioemulsifiers, surface-active agents — is more attributed to the biotechnological impetus than to the physiological meaning. The main properties of biosurfactants are to be described in the following chapters, such as their ability to alter both interfacial and surface tensions and to generate micelles.

The metabolism and function of exolipids are relatively unknown. Furthermore, their function can not necessarily be derived from the amphiphilic structure. The structural differences in comparison to intracellular lipids (mainly phospholipids) are evident. On the one hand, one may sometimes conclude that they are only products of secondary metabolism, on the other hand, a relation to the biosynthesis of cell-wall structures is possible. In any case, the difficulty of recognizing the exact function of such metabolites for the cell is present. Figure 1 summarizes some known conditions which result in the excretion of biosurfactants by microorganisms.

Biosurfactants, to be discussed in this paper, differ to the same degree as to chemically obtained surfactants in comparison to other microbial polymeric amphiphiles, e.g. lipopolysaccharides, lipoteichoic acids, or "emulsan" sometimes being classified as "bioemulsifiers" as well. The latter compounds do not, however, generate micelles.

2 Synthesis and Composition of Biosurfactants

Membrane lipids display a constant qualitative composition; quantitative changes may be effected by growth substrates and by the limitation of certain cations [3]. Membrane lipids are mainly phospholipids; extracellular lipids are predominantly glycolipids. The formation of extracellular lipids depends on both the ratios of C:N and C:P in the medium. Extracellular lipids are synthesized very often during the growth of microorganisms on n-alkanes. The causal relationship of emulsified/solubilized alkane and uptake or utilization seems to be obvious (c.f. 4.1). Excretion in the true biochemical sense is to be discerned from the loss of amphiphilic cell-wall components during fermentation, which is caused by shearing and extraction by substrate hydrocarbons and from products of lysis.

2.1 Microorganisms and Conditions for Synthesis

Various strains of pro- and eukaryotic microorganisms are capable of synthesizing biosurfactants. This ability prevails particularly in some genera (*Actinomycetes*, yeasts of the phyllosphere), as far as it is known. Among the most frequently occurring alkane-utilizing microorganisms the following genera are known as producers of surfactants or emulsifying polymers: *Pseudomonas*, *Acinetobacter*, *Achromobacter*, *Arthrobacter*, *Brevibacterium*, *Corynebacterium*, *Candida*, and *Rhodotorula* [4-10]. The relationship between emulsification/solubilization of alkanes by extracellular lipids and of uptake of alkanes has by now become a valid fact. However, many microorganisms do also synthesize biosurfactants during the growth on water-soluble compounds [11-18]. The extremely high extracellular concentration of lipids during the growth of some microorganisms on alkanes is more likely the result of particular biotechnological process parameters, or of particular conditions of limitation, than a necessary metabolic reaction to alkane utilization [19-21].

Surfactant producing microorganisms may be divided into three groups with respect to alkane utilization and the synthesis of extracellular lipids:

Microorganisms which produce biosurfactants exclusively during growth on alkanes

Corynebacterium sp. [22], *Arthrobacter* sp. [23], *Nocardia* sp. [23], and *Corynebacterium lepus* [24] are representative microorganisms of this group. *Rhodococcus erythropolis* [25] synthesizes trehalose mycolates (Table 1) at an increased rate during growth on alkanes. Only 10% of these lipids are excreted into the culture medium, 90% remain cell-bound. It seems probable that the loss of this small quantity of glycolipid is caused by mechanical stress and/or the influence of hydrocarbons on cells during fermentation.

Mycolic acid and trehalose mycolates are typical components of the cell-walls of many *Actinomycetes*. The hydrophobic cell-boundary of *Actinomycetes* is attributed to these cell-wall components. Many species of the genera *Arthrobacter*, *Brevibacterium*, *Corynebacterium*, *Mycobacterium*, and *Nocardia* are known to degrade hydrophobic compounds, e.g. alkanes, steroides [28,29]. *Candida lipolytica* excretes an amphiphilic component consisting of protein, carbohydrate, and fatty acid only during the late exponential phase of growth on alkanes. This polymer solubilizes hydrocarbons [30,31]. In a similar manner Käppeli [32] was able to detect the function of a cell-bound polymer only during the growth of *Candida tropicalis* on alkanes. Recently Reddy et al. [33] reported on the isolation of a protein-carbohydrate-lipid which contains a compound with emulsifying properties and is excreted during the growth on alkanes by *Pseudomonas* PG 1.

Microorganisms which produce biosurfactants on both alkanes and water-soluble compounds

Microorganisms of the second group represent an extraordinary number of microbial producers of biosurfactants. The best known example is *Pseudomonas aeruginosa*. During the late exponential growth phase this strain synthesizes rhamnolipid (Table 1) with complex media and with synthetic media using glycerol, hexadecane and other carbon sources [34–40]. The synthesis seems to be regulated (inhibited) by some catabolites, e.g. glucose, acetate or citrate [41,42]. In any case, the amount of excreted rhamnolipids depends on the ratios of both C:N and C:P in the medium [43]. The optimum C:N and C:P-ratios in the continuous culture amount to 18 and 10–16, respectively. Among the trace elements, Fe^{++} (as iron (II)sulfate) inhibits the formation of rhamnolipids at concentrations of 2 mg g^{-1} glucose and above. The highest lipid concentration was reached at a C:Fe-ratio of 72400, i.e. 0.5 mg l^{-1} $FeSO_4 \times 7\,H_2O$. Similar tendencies were observed with another strain of *P. aeruginosa* in batch experiments with ammonium iron (II)sulfate and glycerol as a carbon source [44]. In the latter case, the yields during discontinuous cultivation were at least 4 g l^{-1}. The influence of nitrogen-regulation as well as the effect of a limitation of multivalent metal ions on an overproduction of glycolipids is also known from *R. erythropolis* [117].

In addition, resting cells of *P. aeruginosa* are able to synthesize rhamnolipids which are excreted into the medium [44,45]. This has been described for other resting and immobilized microorganisms as well, for instance for *Arthrobacter* sp. and *R. erythropolis* [45].

An extracellular biopolymer with surface-active properties was isolated from *Corynebacterium hydrocarboclastus* cultures during the growth on some alkanes and

Table 1. Structure and origin of biosurfactants

Trivial name	lipophilic moiety	Hydrophilic moiety	Organism	Growth substrate (Remarks)	Ref.
1. Glycolipids Rhamnose lipid	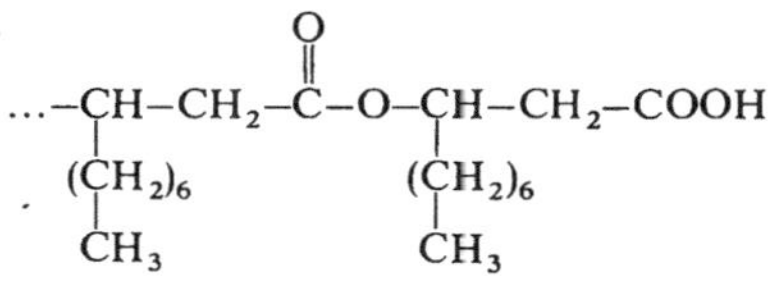	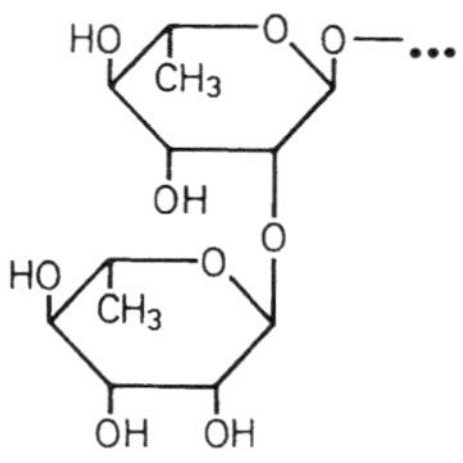	*Pseudomonas aeruginosa*	Glycerol Hexadecane	11, 16, 34, 41, 42, 79)
	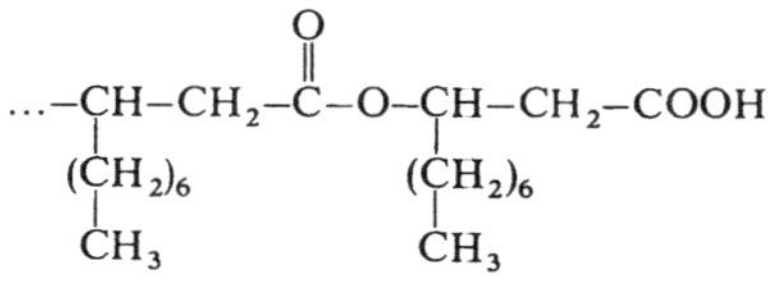	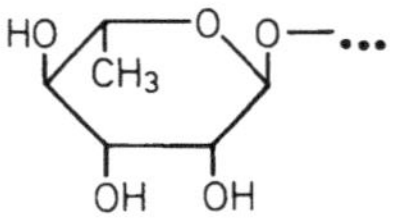	*P. aeruginosa*	Glycerol Hexadecane	34)
	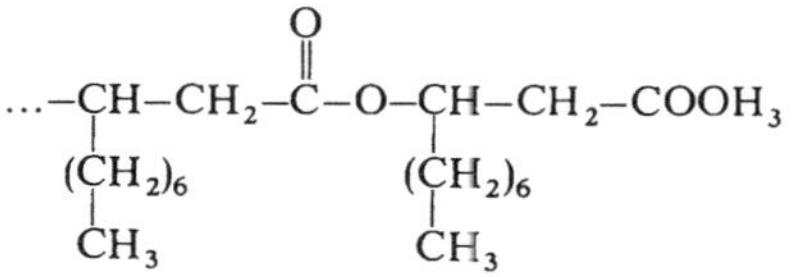 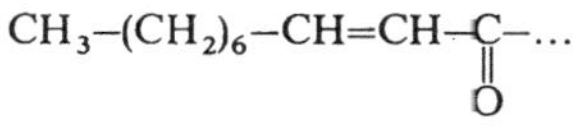	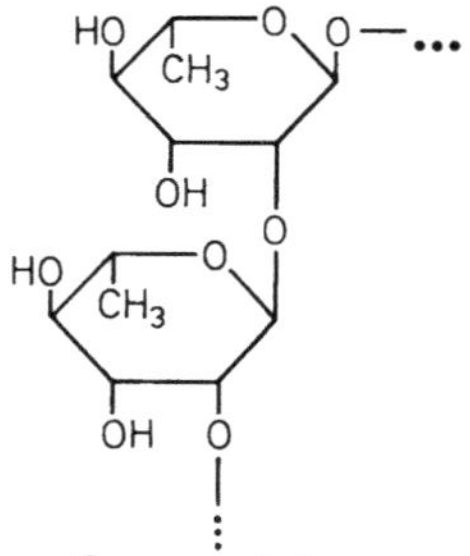 Compound A	*P. aeruginosa*	Hexadecane Glycerol Glucose Ethanol	39)

Table 1. (continued)

Trivial name	lipophilic moiety	Hydrophilic moiety	Organism	Growth substrate (Remarks)	Ref.
	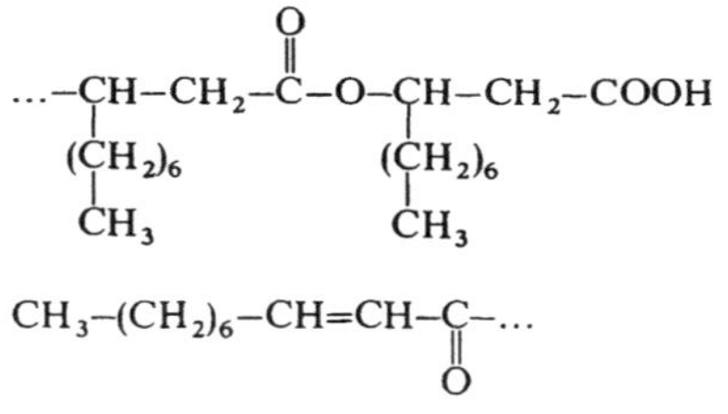	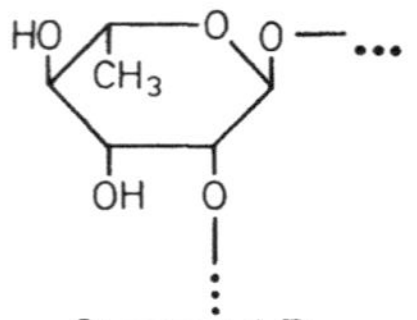 Compound B	*P. aeruginosa* (A:B = 3:1)	Hexadecane Glycerol Glucose Ethanol	[39]
	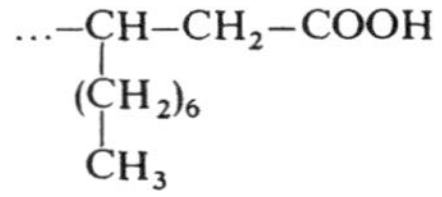	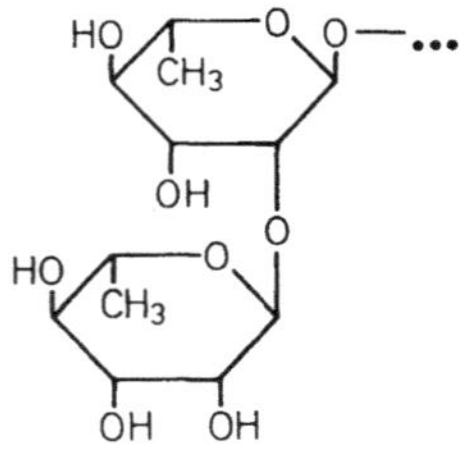	*Pseudomonas* sp.	(Only with resting cells; C-source: n-alkanes, glycerol)	[45]
	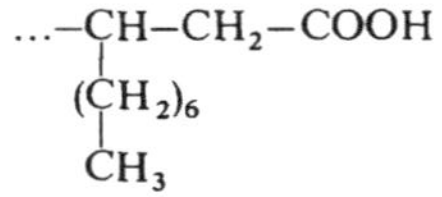	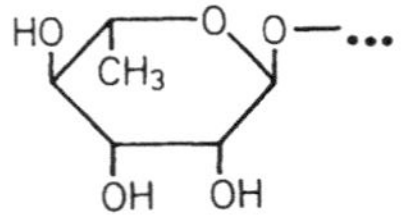	*Pseudomonas* sp.	(Only with resting cells; C-source: n-alkanes, glycerol)	[45]
	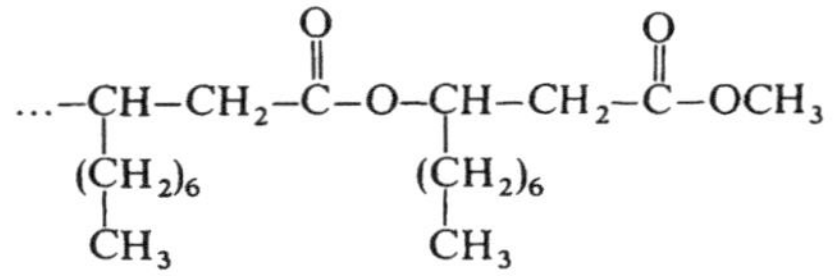	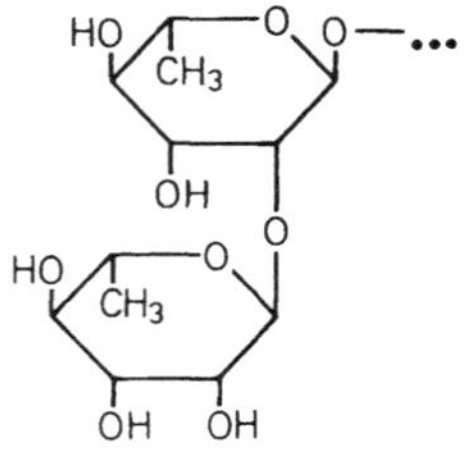	*P. aeruginosa* 158	Trypticase soy broth	[80]

	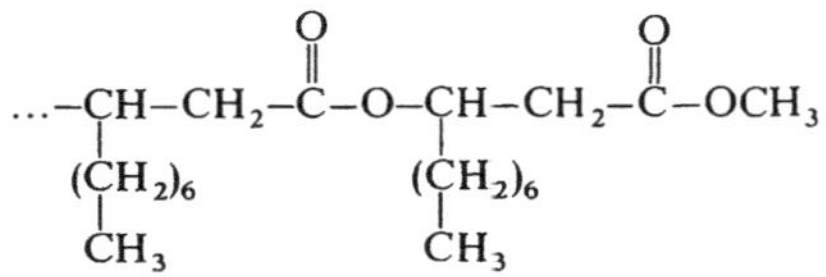	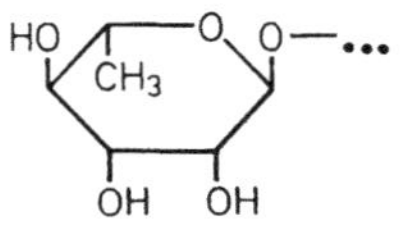	*P. aeruginosa* 158	Trypticase soy broth	80)
Sophorose lipid	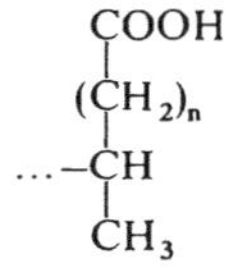 "n" depends on the carbon source, especially on the chain length of the utilized alkane.	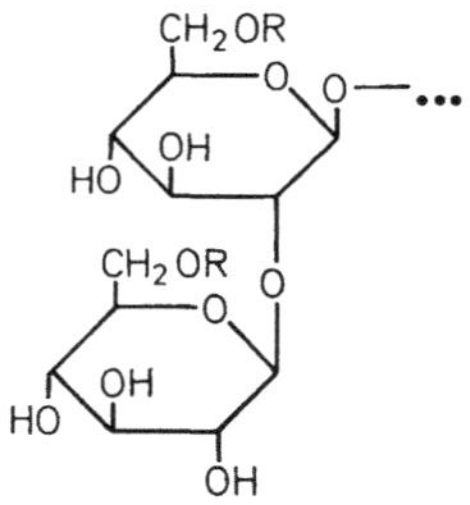	*Torulopsis* sp.	Glucose, Alkane and derivatives	13–15, 81–85)
	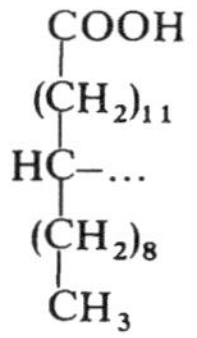	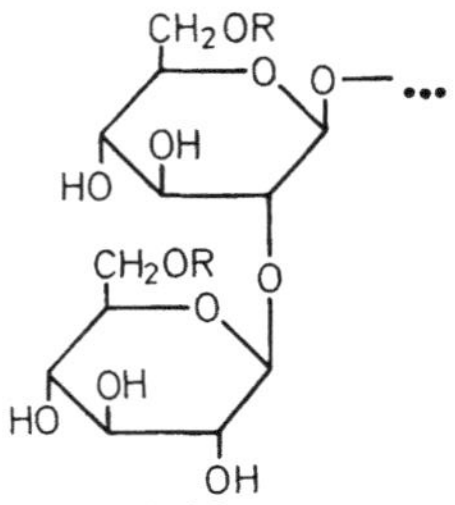 R = CH_3CO	*Candida bogoriensis*	Glucose	86)
Mannosyl-erythritol lipid	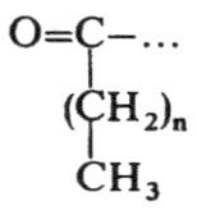 n = 5–12, dependent on the carbon source	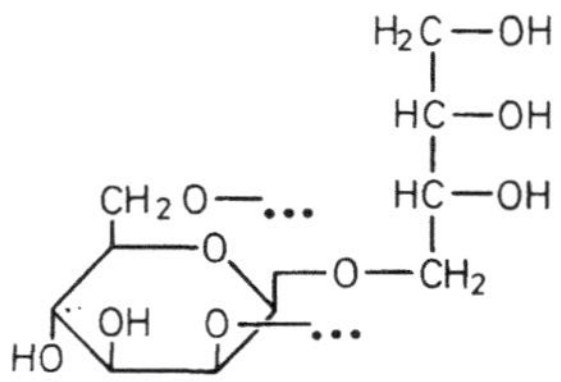	*Candida* sp. (mutant) *Ustilago* sp.	Alkane, Tri-acylglycerol	62, 63)

Table 1. (continued)

Trivial name	lipophilic moiety	Hydrophilic moiety	Organism	Growth substrate (Remarks)	Ref.
Cellobiose lipid	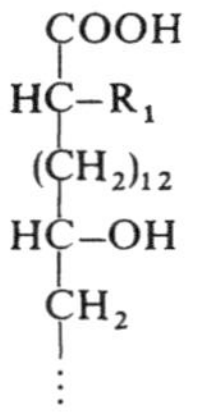 bound to the first glucopyranosylresidue 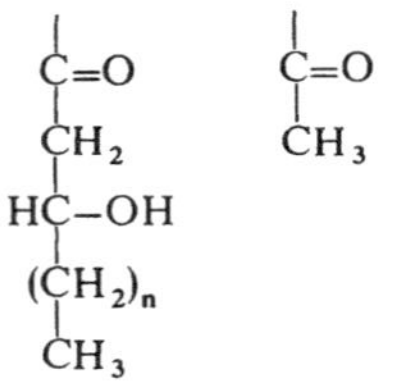 R_1 = H or OH, n = 2–4	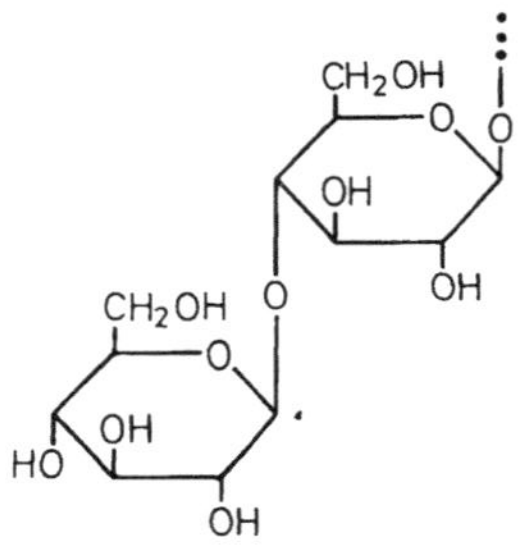	*Ustilago maydis* (Growing and resting cells)	Glucose, Lipophilic substrates	87)
Trehalose mycolate (Trehalose lipid)	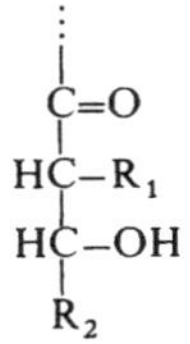 R_1 = alkyl R_2 = alkyl	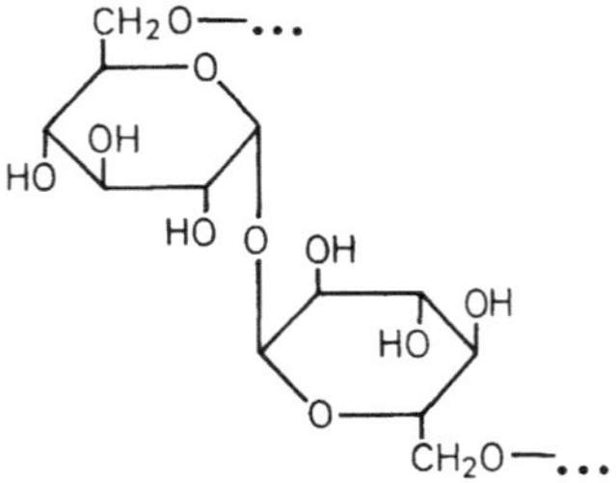	*Mycobacterium* sp. *Brevibacterium* sp. *Arthrobacter* sp. *Corynebacterium* sp. *Nocardia* sp. *Rhodococcus erythropolis*	Alkane	25, 26, 88, 117)
Fructose mycolate	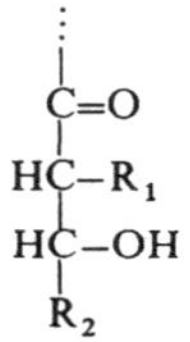 R_1 = alkyl R_2 = alkyl	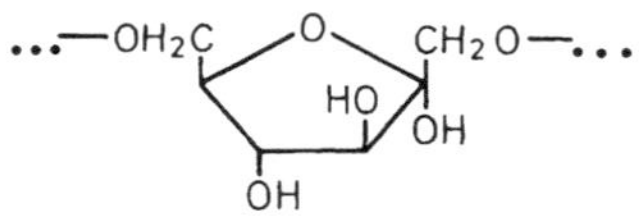	*Arthrobacter* sp. *Corynebacterium* sp. *Mycobacterium* sp. *Nocardia* sp.	Fructose	17, 18)

Sucrose mycolate	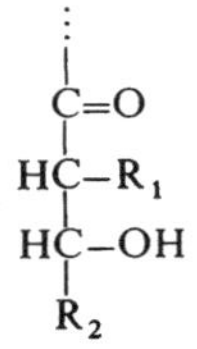 R_1 = alkyl R_2 = alkyl	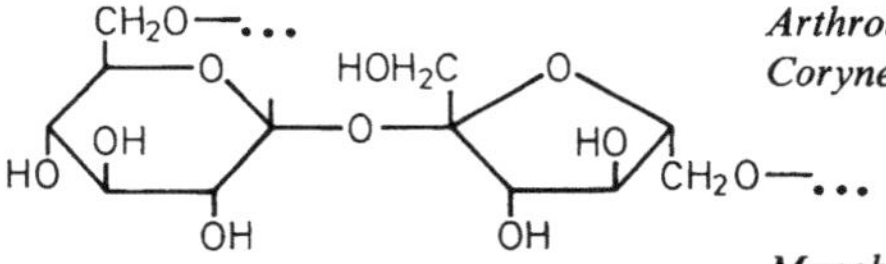	*Arthrobacter* sp. *Corynebacterium* sp. *Mycobacterium* sp. *Nocardia* sp.	Sucrose	[17, 18]
Glucose-6-β-hydroxy-α-hexadecenoyl-eicosenoate	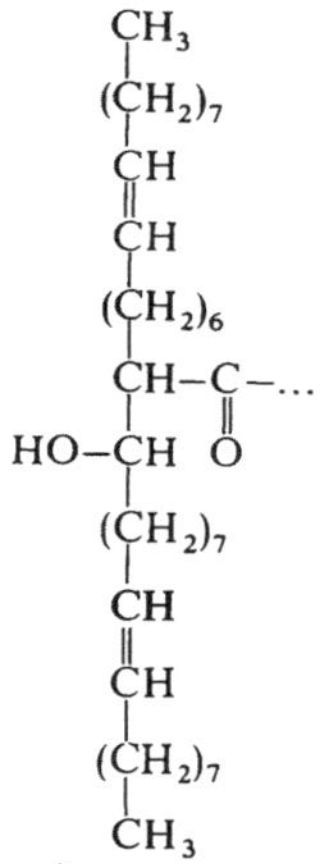	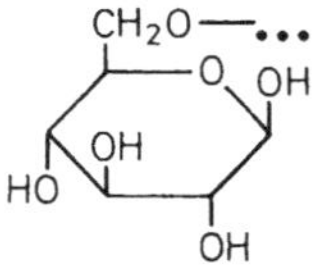	*Brevibacterium thiogenitalis*	(Derived from oleic acid, which is added to the medium)	[64]
Mannitol-β-hydroxypalmitate	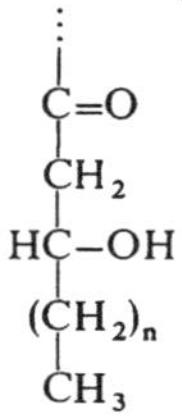	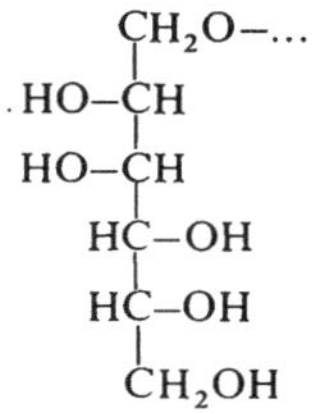 OH-groups are partly acetylated	*Rhodotorula* sp.	Glucose (Also detected: Pentitol-, β-hydroxystearoyl derivatives)	[12]

Table 1. (continued)

Trivial name	lipophilic moiety	Hydrophilic moiety	Organism	Growth substrate (Remarks)	Ref.
2. Amino acid and peptide lipids (cf. MS)					
Ornithine lipid	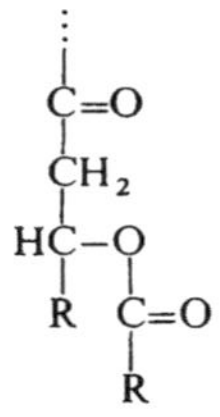 R = alkyl	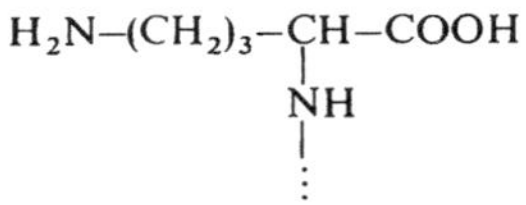	*Pseudomonas rubescens* *Thiobacillus thiooxydans*		66, 89, 90, 91) 89, 92, 93)
Lysin lipid	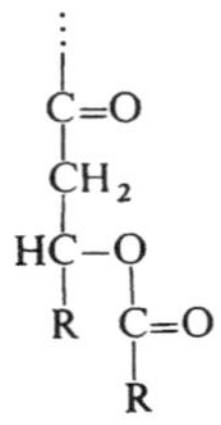 R = alkyl	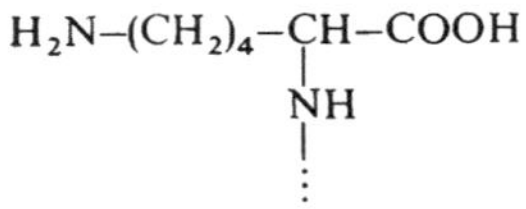	*Agrobacterium tumefaciens*	Complex medium + Glucose	94, 95)
Cerilipine	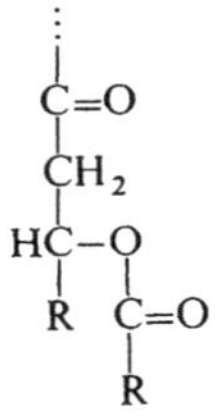 R = alkyl	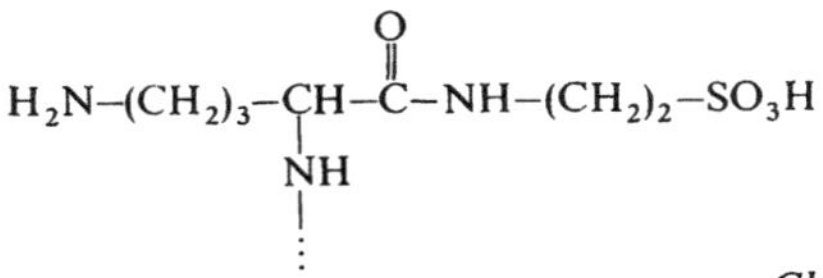	*Gluconobacter cerinus*		91, 95)

Sulfonolipid (e.g. capnine)	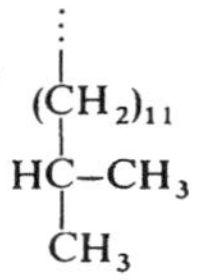 and N-acyl-derivatives	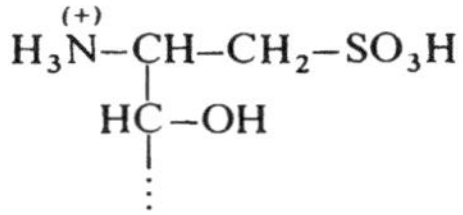	*Capnocytophaga*		96)
Peptide-lipids Surfactin	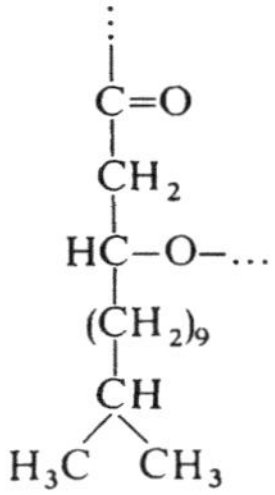	...L-Glu–L-Leu–D-Leu–L-Val ...L-Leu–D-Leu–L-Asp and analogous compounds	*Bacillus subtilis* *B. mesentericus* *Mycobacterium fortuitum* *M. paratuberculosis* *Nocardia asteroides* *Corynebacterium lepus* *Streptomyces canus*	Alkane and other substrates	24, 27, 56–60, 94, 97–99, 100, 257)
3. Fatty acids (cf. MS)					
Mycolic acid	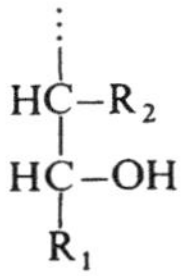 R_1, R_2 = alkyl (varies from genus to genus)	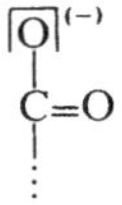	*Mycobacterium* sp. *Corynebacterium* sp. *Nocardia* sp.	Alkane	26) 24, 101) 102)
Long-chain fatty acid	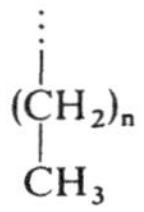	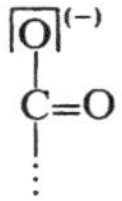	*Bacillus* sp. *Acinetobacter* sp. *Pseudomonas* sp. *Mycococcus* sp. *Penicillium* sp. *Aspergillus* sp.	Alkane	60, 105)

Table 1. (continued)

Trivial name	lipophilic moiety	Hydrophilic moiety	Organism	Growth substrate (Remarks)	Ref.
Dicarboxylic acid	$\vdots$–$(CH_2)_n$–C(=O)–OH	$\lvert\overline{O}\rvert^{(-)}$–C(=O)–$\vdots$	*Candida cloacae* *Pichia* sp. *Corynebacterium*	Alkane	106–111)
4. Glycerol lipids (cf. MS)					
Acyl-glyceride	$\vdots$–C=O–$(CH_2)_n$–CH_3	CH_2O–… CHO–… CH_2O–…	*Acinetobacter*	Alkane	105)
Phospho-lipids	$\vdots$–C=O–$(CH_2)_{10}$–CH_3 and homologous compounds	CH_2O–… CHO–… CH_2O–P(=O)(OH)–O–R	*Cladosporium resinae* *Thiobacillus thiooxydans* *Corynebacterium lepus*	Glucose, Dodecane, Hexadecane	24, 92, 112–116)
Lecithin	R=$CH(OH)$–CH_2–$N(CH_3)_3$		*Corynebacterium alkanolyticum*		
Phosphat-idylethanol-amine	R=$CH(OH)$–CH_2–NH_2				
Phosphat-idylserine	R=$CH(OH)$–$CH(NH_2)$–COOH				
Phosphatidic acid	R = H				

carbohydrates [47,48]. The polymer, or a mixture of polymers, is composed of protein, carbohydrate and lipid. Two critical micelle concentrations (CMC) can be measured in the cell-free supernatant. The yield amounts to about 6.5 g l^{-1} after growth on tridecane, and after growth on glucose, mannitol, or fructose the yield amounts to 1.5 g l^{-1}, 2 g l^{-1} and 2.8 g l^{-1}, respectively. The surface tensions at the second CMC are relatively high compared with those of water ($\gamma_s = 72.8$ mN m^{-1}): 62 mN m^{-1}, 63 mN m^{-1}, 60.5 mN m^{-1} after growth on tridecane, mannitol and fructose, respectively. The emulsifying properties of the polymer (stability of emulsions and diameter of emulsified oil particles) and the interfacial tension versus kerosene may doubtlessly more precisely describe the function of the emulsifier at the interface. In comparison to this, the interfacial tensions of pure solutions of proteins (e.g. serum albumin, serum globulin, ovalbumin, all 10^{-3} g l^{-1} in buffer at pH 3–10) are lowered by 23 mN m^{-1} measured versus *p*-xylene or bromobenzene [49].

Yeasts of the genus *Torulopsis* synthesize sophorose lipids (Table 1) during growth on glucose or on glucose with alkanes or derivatives of these as an additive [13–15,50] and with alkanes as the sole carbon source [51,82], respectively. The chain-length of the fatty acid in the glycolipid is determined by the carbon source and the additive. In any case, excretion begins during the stationary growth phase if nitrogen in the medium is exhausted.

In the case of *Candida* 107 and other oleaginous microorganisms the limitation of growth by nitrogen or another nutrient besides carbon results in *intracellular* lipid accumulation due to a high energy charge [52]. The activity of mitochondrial NAD^+-dependent isocitrate dehydrogenase is low because of a decreased concentration of AMP. Isocitrate is accumulated and is converted to citrate via aconitase which then flows out of the mitochondrion. The cytosolic ATP: citrate-lyase splits citrate into acetyl-CoA and oxaloacetate. Acetyl-CoA is the starting point of the biosynthesis of fatty acids, and moreover, acetyl-CoA carboxylase is activated by citrate. Contrary, the *extracellular* lipid accumulation by some yeasts and the reason for excretion have not been elucidated up to now.

Another extracellular sophorose lipid (Table 1) is synthesized by *Candida bogoriensis* after growth on glucose. The lipophilic part of the amphiphilic molecule is 13-D-hydroxydocosanoic acid. The saccharide is partly acetylated [53].

Microorganisms which exclusively produce biosurfactants during growth on water-soluble compounds

Some species of the yeast *Rhodotorula*, as an example of the third group, produce a mixture of mannitol- and pentitolesters of β-D-hydroxypalmitic acid and β-D-hydroxystearic acid (Table 1) during growth on a complex medium with glucose as a carbon source. The glycolipid is partly acetylated [12].

A very unusual extracellular lipid, namely tetraacetylphytosphingosine, is produced by the yeast *Hansenula cifferi* on YM-agar [54,55]. Differences observed in the yield are correlated to the sex and morphological type of the yeast. In the submerse culture the yield amounts to 150 mg l^{-1}. The reason for excretion is probably a disordered lipid synthesis or the inability to metabolize the derivative in a physiological manner, i.e. by combining with other kinds of molecules. There is no statement about the properties of the lipid as a surfactant, but the known structure displays a typical amphiphilic character.

Itoh and Suzuki [18] detected a modification of the trehalose lipid from *Arthrobacter paraffineus* during growth on sucrose and fructose, respectively. They isolated fructose 6-corynomycolate and fructose 1,6-biscorynomycolate from the cells (Table 1). In the same way, the authors could prove that the sugar moiety of the glycolipid is interchangeable between bacteria of the genera *Brevibacterium*, *Corynebacterium*, *Mycobacterium*, and *Nocardia*. However, in cells of *Mycobacterium* sp. grown on sucrose, sucrose mycolate could not be detected.

The extracellular lipopeptides from some *Bacillus* sp. are very efficient biosurfactants [56, 57]. Cooper et al. [57] isolated the lipopeptide surfactin (Table 1) from the cell-free culture medium after growth of *Bacillus subtilis* on glucose (c.f. also [58–60]). Hexadecane as an additive (20–40 g l^{-1}) inhibited the synthesis or excretion of surfactin although the biomass increased [57]. A high yield of surfactin is caused by adding Mn^{++} (10^{-6} M) to the medium. This is also known for the synthesis of other secondary metabolites [61]. Another secondary metabolite-like lipid is the mannosyl-erythritol lipid (Table 1) synthesized and excreted by a mutant of *Candida* sp. only after growth on n-alkanes and triacylglycerol, respectively [62]. Therefore, this organism should evidently be listed in group 1. However, the structure resembles the glycolipids from *Ustilago maydis* [87], *U. nuda* and *Schizonella melanogramma* [63]. Schizonellin A and B from this last mentioned organism are antibiotics and mainly intracellular metabolites [63].

An oleic acid-requiring mutant of *Brevibacterium thiogenitalis* excretes a glycolipid: 35–45% of the oleic acid is converted to β-hydroxy-α-hexadecenoyl eicosenic acid glycosidically linked with glucose (Table 1) [64].

Pseudomonas rubescens [65], *Agrobacterium tumefaciens* [66] and *Gluconobacter cerinus* [67] synthesize amino acid-containing lipids (Table 1) during growth on water-soluble compounds (glucose). A similar lipid is detected in *Streptomyces sioyaensis*, namely the lysin containing siolipin A [68].

The classification of the cited biosurfactant producers into three groups does not take into consideration the physiological importance of the excreted lipids. In many cases the excretion of lipids is drastically increased in the presence of hydrocarbons: This probably seems to be often only a means of enhancement of the yield.

Käppeli and Finnerty [69, 70] isolated vesicles containing components of the outer membrane from the medium after growth of *Acinetobacter calcoaceticus* on alkanes. The isolation of lipopolysaccharide with strong emulsifying properties from culture filtrates after growth of *A. calcoaceticus* on ethanol is claimed in a Japanese patent [71]. Another strain of *A. calcoaceticus* produces a bioemulsifier after growth on ethanol, hexadecane, and acetate [74]. It is composed of galactosamine, aminouronic acid, glucose and a fatty acid ester with a molecular weight of 9.76×10^5 [72–74].

Another cell surface amphiphile, lipoteichoic acid, is excreted by *Streptococcus faecium* during growth on glycerol [75]. In aqueous solutions these amphiphilic polymers form polydisperse micellar aggregates with particle weights of several million Daltons [76]. In contrast, the micelles of "true" surfactants are in a dynamic equilibrium with the monomers and have a defined size and aggregation number. A criterion for the biological relevance of biosurfactants and amphiphilic polymers during the utilization of hydrophobic substances is the interfacial tension water/apolare phase and apolare phase/cell-boundary, respectively. The system water-apolare phase — biosurfactant, existing above the CMC as a micellar system can pass over to a microemulsion in the presence of electrolytes and a cosurfactant such as alcohols or another surfactant. However, one must not overlook the fact that the physical pressure during cultivation of microorganisms in a homogeneous phase may result in the accumulation of surface-active polymers in the medium caused by shearing or sloughing off of the outer cell structure [77].

Most extracellular lipids are glycolipids corresponding in their composition partially or completely (*Actinomycetes*) to components of the cell wall or to components of the outer membrane. The same situation is shown in the case of the amphiphilic polymers such as lipopolysaccharides, lipoteichoic acids etc.: they are also cell wall related structures. Excretion into the medium reflects biochemical/morphological changes during growth induced by either artificial limitation of components of the medium or a physiological state, i.e., aging of the microorganism. This means, of course, that limitation and a shift in the biosynthetic pathways are also present. The extraction of membrane components by hydrocarbons during microbial utilization of hydrocarbons is well known [78]. These "extractive conditions" may also sometimes be a reason for the occurrence of biosurfactants in the medium [77].

2.2 Structure and Properties

2.2.1 Structure

The structure of the most known biosurfactants is summarized in Table 1. The subdivision of the amphiphilic molecule in a hydrophilic and lipophilic part is related to the chemical structure and not to the real orientation at the interface. As shown in Table 1, the hydrophilic and lipophilic moieties of biosurfactants are diverse. The hydrophilic moieties may be as simple as the carboxylate group of fatty acids or as complex as the mono-, di- and polysaccharides of glycolipid biosurfactants and the polar side chains and peptide backbone of lipopeptide biosurfactants. The lipophilic moieties are saturated, unsaturated or hydroxylated fatty acids. Biosurfactants from *Corynebacterium*, *Mycobacterium* and related microorganisms also contain α-branched β-hydroxymycolic acids with very long chains containing approximately 30 to 90 carbon atoms.

From a chemical point of view, the hydrophilic part of the biosurfactants — responsible for the degree of solubility in water — is bound to the lipophilic part responsible for capillary activity by the:

- Ester linkage (incl. lactones) with organic and anorganic acids
- Amide linkage (single and peptide)
- Glycosidic linkage (sugar-sugar; sugar-hydroxy fatty acid)

Besides this, the ionization of functional groups plays an important role in surface activity, especially if simple carboxylic acids act at the interface water — oil.

In Table 1 only typical growth substrates are listed. The real spectrum of utilization of different substrates and production of exolipids is often much wider. Now and then unusual intracellular lipids such as cerilipine, lysin lipid etc. are also shown to be selectively extractable from the cells.

Table 2. Physico-chemical properties of biosurfactants

Strain	C-source	Type of lipid (cf. Table 1)	Concentration in the medium	γ_s (mN m^{-1})	γ_i (mN m^{-1})	CMC of the pure component	Ref.
Rhodococcus erythropolis	Hydrocarbon	Trehalose 6-mycolate	cell-bound	30	15	165 mg l^{-1} (H_2O)	[25]
		Trehalose 6,6'-dimycolate	cell-bound	37	15	0.7 mg l^{-1} (H_2O)	[25]
Corynebacterium sp.	Kerosene	unknown	0.8 g l^{-1}	40	n.d.	300 mg l^{-1}	[22]
Corynebacterium xerosis	Kerosene	unknown	0.9 g l^{-1}	45	n.d.	500 mg l^{-1}	[22]
Arthrobacter paraffineus NRRL B-3453-1	Nutrient broth + glucose + hexadecane	n.d.[a]	4 × CMC	39	8	n.d.	[23]
A. paraffineus NRRL B-3453-2	Nutrient broth + glucose + hexadecane	n.d.	39 × CMC[d] 10 × CMC	32.5	1	n.d.	[23]
A. paraffineus NRRL B-3453-3	Nutrient broth + glucose + hexadecane	n.d.	17 × VMC	30	1	n.d.	[23]
Corynebacterium fascians ICPB CF16	Nutrient broth + glucose + hexadecane	n.d.	120 × CMC	32	1.5	n.d.	[23]
Nocardia amarae NRRL B-8176	Yeast extract + glucose + hexadecane	n.d.	1000 × CMC	36.5	3	n.d.	[23]
Corynebacterium poinsettiae ICPB CP109	Yeast extract + glucose + hexadecane	n.d.	4 × CMC	30	6	n.d.	[23]
C. fascians ICPB CF15 matt	Yeast extract + hexadecane	n.d.	16000 × CMC[d] 160 × CMC	28.5	1.5	n.d.	[23]
C. fascians ICPB CF15 non-matt	Yeast extract + hexadecane	n.d.	213 × CMC[d] 190 × CMC	29.5	2	n.d.	[23]
C. fascians ICPB CF16	Yeast extract + hexadecane	n.d.	10 × CMC	33	2	n.d.	[23]
C. fascians ICPB CF17	Yeast extract + hexadecane	n.d.	6250 × CMC[d] 45 × CMC	27.5	1.5	n.d.	[23]
C. fascians ICPB CF21 matt	Yeast extract + hexadecane	n.d.	180 × CMC	31	4	n.d.	[23]

C. fascians ATCC 12974	Yeast extract + hexadecane	n.d.	30 × CMC	30	2	n.d.	[23]
C. fascians NRRL B-190	Yeast extract + hexadecane	n.d.	670 × CMC	33	2	n.d.	[23]
Corynebacterium insidiosum ICPB CI 13A	Yeast extract + hexadecane	n.d.	25 × CMC	28.5	0.55	n.d.	[23]
Corynebacterium michiganense ICPB CM 179A	Yeast extract + hexadecane	n.d.	4 × CMC	60.5	28	n.d.	[23]
Corynebacterium lepus	Kerosene	Corynomycolic acid Lipopeptides a.o.	60 × CMC	30	2	150 mg l^{-1}	[24, 118]
Pseudomonas asphaltenicus	Kerosene	Fatty acid + β-hydroxy fatty acids	<1 × CMC	30	15 (for crude prod.)	200 mg l^{-1}	[118]
Corynebacterium hydrocarboclastus	Kerosene	Polymer (protein-lipid-carbohydrate)	n.d.	49/35	15/2	1.7 g $l^{-1\,d}$ 4.7 g l^{-1}	[48]
C. hydrocarboclastus	Fructose	Polymer (protein-lipid-carbohydrate)	3 g l^{-1}	68/60.5	n.d.	2.2 g $l^{-1\,d}$ 8.1 g l^{-1}	[47]
	Heptadecane	Polymer (protein-lipid-carbohydrate)	3.5 g l^{-1}	55/39.5	n.d.	2.4 g $l^{-1\,d}$ 6.3 g l^{-1}	[47]
Bacillus subtilis	Glucose	Lipopeptide	0.1–0.8 g l^{-1}	27	1	25 mg l^{-1}	[57]
Candida lipolytica	Alkane	Polymer (protein-lipid-carbohydrate)	n.d.	44	n.d.	5 g l^{-1}	[30]
Nocardia erythropolis	Hexadecane	Fatty acid, mixture of neutral lipids	160 × CMC	32	2	n.d.	[20]
C. fascians	Kerosene	Corynomycolic acid	75 × CMC	33	n.d.	2 g l^{-1}	[19]
C. hydrocarboclastus	Kerosene	Corynomycolic acid	8 × CMC	32	n.d.	2 g l^{-1}	[19]
C. lepus	Kerosene	Corynomycolic acid	15 × CMC	30	n.d.	150 mg l^{-1}	[19]
C. xerosis	Kerosene	Corynomycolic acid	7 × CMC	29	n.d.	1 g l^{-1}	[19]
Pseudomonas spec.							
— growing cells	Alkane	Rhamnolipids	180 mg g^{-1} substrate	26	<1	10 mg $l^{-1\,b}$	[45]
— resting cells	Alkane, Glycerol	Rhamnolipids	230 mg g^{-1} substrate	25	<1	200 mg $l^{-1\,b}$	[45]

Table 2. (continued)

Strain	C-source	Type of lipid (cf. Table 1)	Concentration in the medium	γ_s ($mN\ m^{-1}$)	γ_i ($mN\ m^{-1}$)	CMC of the pure component	Ref.
— immobilized cells	Glycerol	Rhamnolipids	110 $mg\ g^{-1}$ substrate	25	<1	200 $mg\ l^{-1}$ [b]	45)
Rhodococcus erythropolis		Trehalose coryno-mycolate/Trehalo-setetraester					
— growing cells	Alkane		280 $mg\ g^{-1}$ substrate	32/26	16/<1	2 $mg\ l^{-1}$/ 10 $mg\ l^{-1}$ [b]	45)
— resting cells	Alkane		650 $mg\ g^{-1}$ substrate	32/26	16/<1	2 $mg\ l^{-1}$/ 10 $mg\ l^{-1}$ [b]	4
Ustilago maydis		Cellobiose lipid					
— growing cells	Glucose		8.6 $mM\ l^{-1}$ Cellobiose	30	<1	20 $mg\ l^{-1}$ [c]	87)
— resting cells	Coconut oil		14 $mM\ l^{-1}$ Cellobiose	30	<1	20 $mg\ l^{-1}$	87)

[a] n.d. = not determined;
[b] γ_s, γ_1, and CMC: measured in synthetic deposit water at 40°;
[c] γ_s, γ_i, and CMC: measured in synthetic deposite water;
[d] Two values of both "concentration in the medium" and "the CMC of the isolated pure components" indicate the generation of two biosurfactants

2.2.2 Properties

The properties of biosurfactants are described as those of synthetic detergents with physico-chemical data: CMC, γ_s, γ_i, type of emulsion and its stability, PIT, and, additionally, in some cases, the alteration of CMC by high concentrations of electrolytes, pH and temperature.

One of the most common parameters to correlate structure and interfacial functionality and the CMC is the hydrophile-lipophile balance (HLB). Surfactants which are more hydrophobic than hydrophilic have HLB values less than 7 whereas, predominantly hydrophilic surfactants have HLB values greater than 7. All known biosurfactants range over the whole HLB scale: the w/o-type (mycolic acid and trehalose dimycolate, HLB values less than 7) and the o/w-type (rhamnolipid, HLB values greater than 7) [77]. Several values of surface tension (γ_s) and interfacial tension (γ_i) measured during the microbial growth on alkanes are extremely low (<30 mN m^{-1} and <1 mN m^{-1}, respectively), so that one may anticipate the formation of microemulsions, if a cosurfactant and a certain concentration of electrolytes are present.

In Table 2 the physico-chemical data of some biosurfactants are summarized. However, in many publications extracellular lipids are described from a pure biochemical or biological point of view although they are evidently potential biosurfactants which can be deducted from the amphiphilic structure. For these compounds all physico-chemical data fail and, therefore, they are not included in the table.

The increasing interest in biosurfactants in the field of biotechnology is reflected by the increasing number of patents during the last few years (Table 3) and also aims at the "latent" biosurfactants.

3 Biosynthesis of Biosurfactants

In contrast to the increasing number of papers describing microorganisms producing biosurfactants as well as their properties and, thereafter, aspects of application, only a small number of authors have attempted to clarify the metabolic sequences of biosurfactant biosynthesis. Up to now, nearly all proposals to improve yields have been reduced to phenomenological work. However, it has become an essential prerequisite to study the pathways of synthesis in question in order to devise those cultivation conditions, including the use of regulatory mechanisms, for large scale biosurfactant production. The connections between the fatty acid metabolism on the one hand and the pathways responsible for the hydrophobic moiety of the amphiphilic molecule on the other are for the most part unknown.

3.1 Acylated Sugar Derivatives

Glycolipid biosurfactants are acylated sugar derivatives which do not contain glycerol. Investigations on the biosynthesis were accomplished by Hauser et al. [41,42] and Burger et al. [40] with rhamnolipid, by Jones and Howe [81-85] with sophorolipid, by Kretschmer and Wagner [102] with trehalose lipid, and by Lederer [26] with mycolic acid and trehalose mycolate. Because of the different structure of the mycolic acid derivatives their biosynthesis is discussed separately.

Table 3. Claimed inventions of emulsifying agents produced by microbiological processes

Strain	C-Source	Type of surfactant	Yield	γ_s ($mN\ m^{-1}$)	γ_i ($mN\ m^{-1}$)	Ref.
Corynebacterium hydrocarboclastus	Kerosene	Polymer	5–6 $g\ l^{-1}$	n.d.[b]	14.6	119)
Arthrobacter sp.	Alkane	Fatty acid	n.d.	n.d.	n.d.	120)
Brevibacterium ketoglutamicum	Cycloparaffine	esters of glucose, trehalose or sucrose				120)
Micrococcus paraffinolyticus	Aromatic compounds					120)
Corynebacterium sp.						120)
Mycobacterium smegmatis	Kerosene					120)
Candida lipolytica						120)
Aspergillus oryzae						120)
Nocardia rhodochrous	Alkane	Glycolipid (cell-bound)	0.6 $g\ l^{-1}$	n.d.	n.d.	121)
Corynebacterium sp.	Kerosene	Mixture of phospholipids, glycerides	30000 × CMC[a]	29	1	122)
Nocardia sp.	Hydrocarbon	Fatty acid/ester, unknown components	150 × CMC	n.d.	n.d.	122)
Torulopsis apicola	Glucose, with alkane and derivatives as additives	Fatty acid esters of sophorose	Conversion rate of 80 %	n.d.	n.d.	123)
Torulopsis apicola H157	Glucose, Alkane	Fatty acid esters of carbohydrate	> 1.500 × CMC	22–35[c] (depends on C-source)	0.9	51)
Torulopsis glabrata H159	Glucose, Alkane		2.1 × CMC	39[c]	32	51)
Torulopsis candida Y 127	Alkane		n.d.	50[c]	n.d.	51)
Torulopsis famata H158	Alkane		n.d.	50[c]	n.d.	51)
Torulopsis magnoliae	Glucose, with alkane and derivatives as additive	Fatty acid esters of sophorose	Conversion rate of 45–60 %	n.d.	n.d.	124)
Pseudomonas aeruginosa	Glucose, Glycerol	Fatty acid esters of	4 $g\ l^{-1}$	27[c]	0.25	44)

196 Aa		rhamnose				
Mycobacterium sp.	Alkane	Fatty acid ester of trehalose	1.4 g l^{-1}	n.d.	n.d.	[125]
Micrococcus paraffinolyticus	Alkane	Fatty acid ester of trehalose	2.2 g l^{-1}	n.d.	n.d.	[126]
Arthrobacter paraffineus	Alkane	Glucosemono- and Trehalose-diester of α-alkyl-β-hydroxy-fatty-acid	2.1 g l^{-1}	n.d.	n.d.	[127]

[a] Factor × CMC;
[b] n.d. = not described;
[c] γ_s of the cell-free culture medium

3.1.1 Glycolipid Derivatives

The biosynthesis of rhamnolipid by *P. aeruginosa* was investigated by estimation of the metabolic intermediates [40]. Thereafter, a reaction scheme as shown in Fig. 2 is proposed. The precursor of rhamnolipid — the activated sugar thymidine diphosphate α-L-rhamnose — derived after stepwise reduction of TDP-α-D-glucose — is sequentially transferred to β-hydroxydecanoyl-β-hydroxydecanoate (reaction II and III). The two rhamnosyl derivatives — mono- and dirhamnosyl glycoside — can be detected in the cell-free medium [127]. It was postulated that the monorhamnosyl derivative is the precursor of the dirhamnosyl one [16]. The hydroxy group of the hydroxydecanoic acid is in D-configuration.

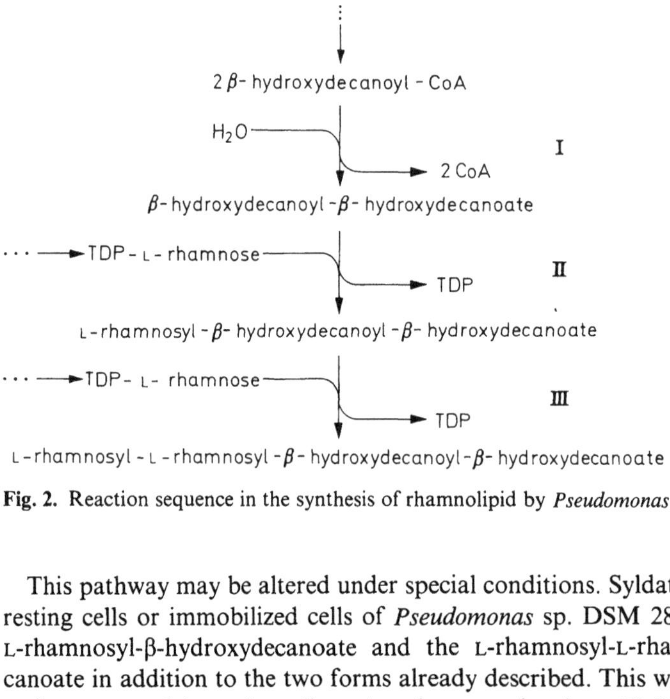

Fig. 2. Reaction sequence in the synthesis of rhamnolipid by *Pseudomonas aeruginosa* [40]

This pathway may be altered under special conditions. Syldatk et al. [45] found that resting cells or immobilized cells of *Pseudomonas* sp. DSM 2874 generate both the L-rhamnosyl-β-hydroxydecanoate and the L-rhamnosyl-L-rhamnosyl-β-hydroxydecanoate in addition to the two forms already described. This was independent of the carbon source (glycerol, n-alkanes) and occurred only at 30 °C not at 37 °C. These slightly altered structures indicate both a uniform metabolic sequence and a modified reaction in the last step (reaction I in Fig. 2), possibly caused by an unknown regulatory feature of the metabolism of resting cells.

The biosynthesis of the sophoroside in *Candida bogoriensis* covers the glucosylation of 13-L-hydroxydocosanoic acid followed by a second glucosylation and two acylation reactions in which two glucosyl- and an acetyl transferase should be involved [128, 129].

When investigating the n-alkane hydroxylating system of *T. gropengiesseri*, Jones and Howe [81, 82] found that oxidation products of alkanes and its derivatives are into sophorosides. The main constituents of the fatty acid moiety of glycolipids generated from C_{12}–C_{24} alkanes were identified as oxygenated alkane derivatives. Their chain length depends on that of the converted alkane. The hydroxylation occurred in the ω and ω-1 position. Therefore, the appropriate hydroxyalkanoic acids have been detected in the glycolipids. The nature of the fatty acid moiety is also influenced by the addition of methyl esters or other compounds [16].

Torulopsis yeasts grown on sugar, usually glucose, seem to metabolize the long-chain alkanes supplied, or their derivatives, without considerable modification of the lipid moiety of sophorolipid. Recently, Hommel et al. [51] also described a glycolipid generation by a *Torulopsis* strain growing on either carbohydrates or polyols. Thereafter, a de novo synthesis of the lipid moiety may also occur [15, 51].

3.1.2 Mycolic Acid Derivatives

The best known mycolic acid derivative is the "cord factor". The synthesis of the precursor trehalose 6-phosphate has been investigated in *Mycobacterium tuberculosis* [130]. A trehalose 6-phosphate synthetase [131, 132] catalyzes the formation of trehalose-6-phosphate by linkage of two α-D-glycopyranosyl units at C-1 and C-1′ according to the following reaction:

$$\text{UDP-glucose} + \text{glucose 6-phosphate} \rightarrow \text{trehalose 6-phosphate} + \text{UDP}.$$

The reaction product is usually hydrolyzed by trehalose 6-phosphate phosphatase at such a rate [132] as to exclude accumulation of phosphorylated sugar which is in general the precursor of the "cord factor".

In contrast, the pathway of the synthesis of the mycolic acid residue is very unclear. Inhibition studies on the fatty acid synthetase or on the fatty acid degradation indicate that mycolic acids are mainly derived from higher fatty acids by chain elongation rather than directly from the acetate pool of cells [133].

Recently, Kretschmer and Wagner [102] investigated the biosynthetic intermediates of the trehalose dicorynomycolates isolated from *R. erythropolis* and postulated a reaction sequence which differs in some aspects to results obtained from other *Actinomycetes*. In *R. erythropolis* both the lipid and the carbohydrate moiety are separately formed and subsequently gradually linked step by step by esterification (Fig. 3). Contrary to the pathway of the sugar moiety of the "cord factor", the trehalose 6-phosphate — not the trehalose — is condensed to the appropriate sugar corynomycolate. This reaction seems to be limiting the biosynthetic flux. Thereafter, the addition of precursors to the growing culture, i.e. glucose and free corynomycolic acids, provides an improved biosynthetic productivity [134]. The biosynthesis in question is induced by n-alkanes. It could be demonstrated that the fatty acid moiety of the lipid is generated from the inducing growth substrate.

From other *Actinomycetes*, e.g. *Arthrobacter*, *Corynebacteria*, *Mycobacteria*, *Nocardia* [18, 135, 136] and *Brevibacterium thiogenitalis* [64] other intermediates such as monosaccharide acylates, glucose corynomycolates, α,α′-trehalose-6-(3-oxo-2-tetradecanyl) octadecanoate and/or α-α′-trehalose-6-mycolate-6-acetate of glycomycolate biosynthesis have been detected [64, 135–138]. This indicates, that in other genera the synthesis of mycolates may occur via other sequences which differ with respect to the status of the generation of the first ester linkage, i.e., the introduction of the respective sugar components into the lipid synthesis and with respect to the kind of the sugar as well as to the sequence of esterification.

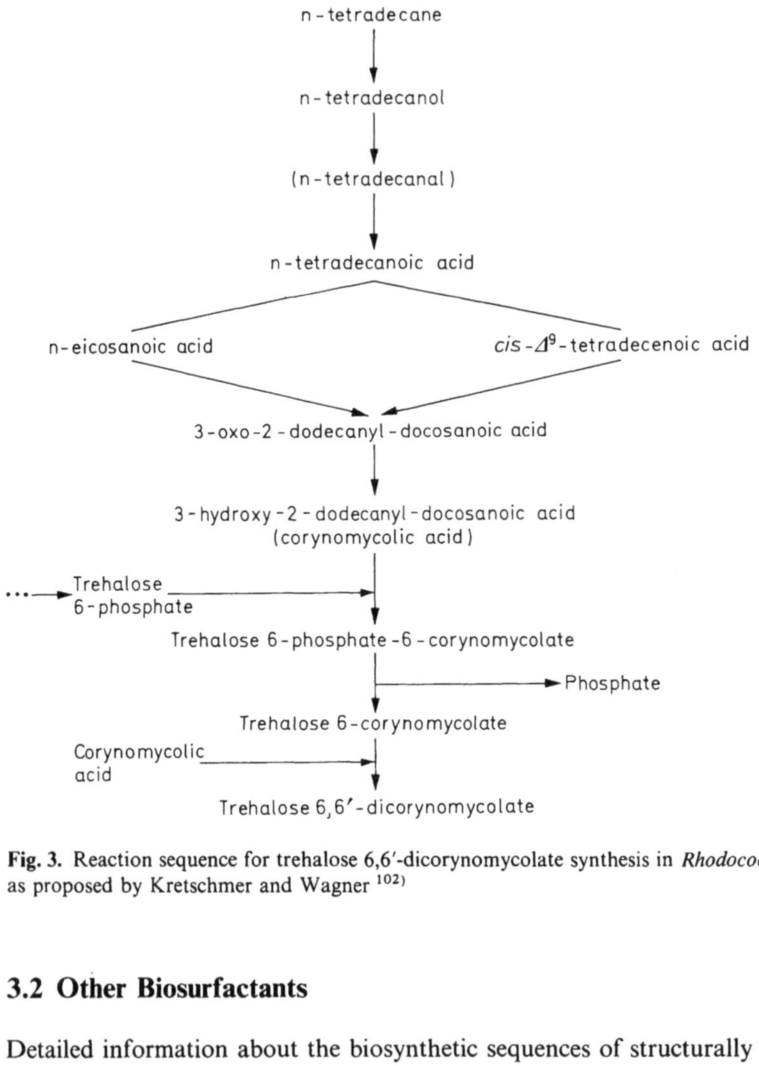

Fig. 3. Reaction sequence for trehalose 6,6'-dicorynomycolate synthesis in *Rhodococcus erythropolis* as proposed by Kretschmer and Wagner [102]

3.2 Other Biosurfactants

Detailed information about the biosynthetic sequences of structurally different biosurfactants, e.g. peptide lipids is not available. In comparison, the pathway of fatty acid synthesis, e.g. [139] as well as the synthesis of peptides or glycerolipids have been well investigated. In this case, the appropriate reviews should be referred to.

4 Physiological Importance of Biosurfactants

All reviewers dealing with biosurfactants have reduced this matter simply to a participation in assimilation of n-alkanes by yeasts and bacteria (e.g. refer [254, 255]). Nevertheless, their exact physiological function remains uncertain in most cases. Some microbial surface-active substances are involved in bacterial gliding [96, 140]. Phospholipids excreted by *Thiobacilli* play an essential role in the wetting of sulphur. Other biosurfactants exhibit biocidic activities. In contrast to glycolipids from mammalians, for instance, the recent information pertaining to biosurfactants is very limited. The former lipid class (together with glycoproteins) has been associated with such phenomena as cell adhesion, recognition, differentiation, and tumorigenic degradation [142].

4.1 Biosurfactants and Hydrocarbon Assimilation

The cultivation of microorganisms on hydrocarbons introduces a new phase in the fermentation system in the form of the almost water-insoluble carbon source, in addition to cells, the gas-phase and the culture medium. Microorganisms have to adapt to these altered conditions. In addition to the limited mass transfer, the solubility of oxygen in the oil phase is greater than in the water [143]. The interfacial conditions existing between gases, water, and cells are disturbed by changes in the surface tension [144].

4.1.1 Theoretical Considerations and Practical Relevance

Three hypotheses have been discussed in the literature in order to understand the availability and uptake of alkanes which has been the subject of considerable study in recent years. These will briefly be considered here (for reviews refer to [145–148]).

— *Uptake of monodisperse-dissolved alkanes* [149]

Table 4 shows the solubility of alkanes in water and the alteration of this amount by growing *Endomycopsis lipolytica* (*Saccharomyces lipolytica*) [150]. The respective amount of dissolved alkanes does not explain the observed specific growth rates of microorganisms on alkanes [143–154].

— *Direct contact of cells with large oil drops* [155–159]

The hydrophobic properties of the outer cell surrounding structures have been made responsible for an adhesion to hydrocarbons. Recently, Rosenberg et al. [141, 160] and Stuber [161] have shown that most of those strains which are very good alkane utilizers possess a strong hydrophobic outer membrane in contrast to less effective or non-hydrocarbon utilizing strains. These surface properties favour these microorganisms during hydrocarbon oxidation by influencing adhesion to the surface of the carbon source [162–164]. However, this property is not only restricted towards alkane substrates or towards alkane-degrading microorganisms [160, 165]. Differences in hydrophobicity of the cell surface during growth [165] indicate a larger influence of the physiological state of the cells than that of the corresponding carbon source. Alterations of the composition of the outer membrane, e.g. an increased incorporation of fatty acids [166, 167] leading to a higher cell hydrophobicity, more strongly marked regions [166, 168, 169], the occurrence of fimbria and membrane vesicles [170–174] accompany microbial growth on alkanes as well.

It is obvious that one way of adaptation may be the generation of cellular metabolites localized at the outer cell wall and which exhibit surface active properties enabling a facilitated transport of the hydrocarbons into the cell.

— *Contact with fine oil droplets (accomodated oil)* [173, 175]

The generation and excretion of cellular metabolites with surface active properties is the basis of this hypothesis.

This model is supported by the findings that emulsified alkanes are much faster assimilated by cells [156, 176–179] and that low detergent concentrations improve growth rates of microorganisms on hydrocarbons [164, 173, 180–182].

The degree of emulsification influences the size of surface areas for contact between cells and alkanes and a quasi solubility of substrate in the water phase. Thereafter, a basal connection may be established between emulsification and solubilization of alkanes on the one hand and influx, utilization, and growth, respectively, on the other hand. Roy et al. [150] was able to show that both emulsifying and solubilizing factors are involved in alkane uptake by *E. lipolytica* (Table 4). The biosynthesis of the appropriate factors is independent of the netto synthesis of extracellular protein but correlates to the cell growth. In spite of different physical states of the molecular dispersed and the solubilized alkanes, respectively, concerning the transport into the cell, both "types" of alkanes act in a similar manner [183].

Table 4. Solubilization of alkanes during growth of *Endomycopsis lipolytica* Y13 [150] compared with those in water. The solubility of different alkanes were taken from [229, 230]

Carbon source	Solubility in water ($g\ l^{-1}$)	Max. solubilization during growth ($g\ l^{-1}$)	Increase of pseudo-solubility (-fold)
Decane	5.2×10^{-5}	0.5	10^4
Dodecane	8×10^{-6}	0.6	7.5×10^4
Tridecane	8×10^{-6}	0.56	7.0×10^4
Tetradecane	7×10^{-6}	0.35	5.0×10^4
Pentadecane	6×10^{-6}	0.6	10^5
Hexadecane	6×10^{-6}	0.4	6.6×10^4
Heptadecane	6×10^{-6}	0.8	1.3×10^5
Octadecane	6×10^{-6}	0.6	10^5
Nonadecane	6×10^{-6}	0.5	8.0×10^4

4.1.2 Effect of Surface-Active Agents on Hydrocarbon Assimilation

The conclusion that surface-active agents are involved in alkane degradation by microorganisms is supported by the generation of appropriate compounds during such processes in many cases (cf. 2.1 and 2.2 and also [16]) as well as by the effects of both biogenic and synthetic tensides [150, 182, 184–186] on the duration of various phases of growth and growth rates. Large differences have been reported on stimulation and/or inhibition by a variety of surfactants. Tanaka and Fukui [180] related the effect of surfactants with their HLB-values. In the early growth phase of *Candida* sp. lipids with low HLB-values exhibited stimulatory effects. In the late growth phase, lipids with higher HLB-amounts were most efficient. Contrary to this, Whitworth et al. [182] have shown that the structure of the acyl-chain may be dominant. Furthermore, Nakahara et al. [185] have shown an influence of Tween 20 on growth of *Pseudomonas* sp. only on hexadecane. Other bacteria and yeasts were either inhibited or not influenced.

These examples demonstrate the lack of knowledge of the real mechanisms as to how surfactants act on hydrocarbon assimilation. Probably the stimulatory or inhibitory effects depend on whether the hydrophobic part of the surfactant may interact with the alkane droplets alone or only with both the droplets and the cells [182]. However, the surface properties of the cells themselves should not be forgotten in this connection, as mentioned above.

The effects of biosurfactants on alkane assimilation are similar. Hisatsuka et al. [36, 37] showed a synergistic effect of rhamnolipid from *P. aeruginosa* and a protein-like activator on growth and oxidation of alkanes by this strain. The presence of these factors was essential for growth [187]. A rhamnolipid-negative mutant possessing the appropriate enzyme system was able to grow on alkanes only after addition of rhamnolipid [34]. The purified rhamnolipid stimulated the growth of *P. aeruginosa* whereas other bacteria were only slightly or not effected. No influence was detectable with a glucose medium [36]. Similar results have been reported by Koronelli et al. [188]. Their investigations showed that *P. aeruginosa* P-20 released a peptidoglycolipid into the medium. In the case of *Pseudomonas cepacia* the formation of emulsifying agents during growth on 2,4,5-trichlorophenoxyacetic acid is known and it has been postulated that these agents have functions in transport of the growth substrate into the cells [189].

Sophorolipids produced by *Torulopsis bombicola* caused a remarkable increase in biomass up to 30% on alkanes of the producer itself [190]. Furthermore, this yeast was enabled to grow very well on the alkanes decane to eicosane not usually being utilized. A number of synthetic surfactants were not able to replace this glycolipid [191]. The effect of this lipid was almost restricted to *Torulopsis* yeasts [191, 192].

The mannosylerythritol lipid from *Candida* sp. [62, 185, 186] improved the productivity of the mutant and exhibited the same stimulation of growth as with "Span 80". Such effects are also known for a peptid-lipid of a hydrocarbon-assimilating yeast [193].

As described previously, the biosurfactants mentioned possess a highly specific influence on different microorganisms growing on alkanes. The most potent activation was always obtained with the producers themselves. Other bacteria or yeasts were either inhibited or not effected. This may be an evolutionary advantage.

The examples described conform in general to the effects reported for synthetic tensides. However, sometimes a stimulation of other microorganisms was observed when an isolated biosurfactant was supplemented to growth medium [161, 191]. This indicates the presence of more complex mechanisms of alkane-cell-surfactant interactions and thus more information is required about cells, their surface layers, and biosurfactants. The most important factor seems to be the cell surface. In this connection it is of interest that cells of *T. bombicola* [190] and *P. aeruginosa*, respectively, have no binding affinity to hexadecane during growth on the alkane in the presence of the appropriate biosurfactant. More complex alkane emulsification and to a certain extent also solubilizing polymers have been described for yeasts and bacteria which show binding affinities for n-alkanes. In most cases these are characterized as being excreted complexes from the cell surface either on the basis of metabolic reactions or by competing reactions with extracellular apolar substrates or amphiphiles, respectively [30, 69, 73, 74, 167, 194, 195, 253]. The emulsifying complex from *E. lipolytica* appeared to be specific for the carbon source [150].

Pareilleux [30] detected a concentration-dependent inhibition of the microorganism by which the emulsifying complex was produced.

In addition to the discussed role of biosurfactants in facilitating the uptake of alkane by the cells, another point of view should be noted. In case of *P. cepacia* Banerjee et al. [189] discussed a more general function in acting as a barrier between the toxic growth substrate present in the medium and the cell membrane resulting in a limited transport of the carbon source into the cell. Therefore, low non-toxic concentrations were maintained inside the cell thus permitting rapid sustained growth. Similar observations had been made by Whitworth et al. [182]. But here the attachment of surfactants to the cells resulted in the formation of a "cell micelle" (consisting of the cells) leading to an inhibition of cell growth.

Generally, such effects described for *P. cepacia* cannot be ruled out when investigating the function of biosurfactants in alkane utilization. Perhaps these circumstances are also a possible explanation for inhibitory activities of certain synthetic and biological surfactants.

Investigations of biosurfactant production in *Corynebacteria* and others indicated alterations of the affinity of the cell surface layers during growth on water insoluble alkanes [196]. This phenomenon may be due to the different behavior of the biosurfactants which may oscillate between an attached and a free soluble or micellized form. Wagner et al. [25] obtained only 10% of the glycolipid of *R. erythropolis* from the supernatant of the culture broth. The bulk amount of the glycolipids [197, 198] remains bound to the cells covering the surface. This may facilitate the diffusion [25] of hydrocarbons into cells. Ignoring the 10% dissolved in the aqueous medium, the second model for alkane uptake, as discussed above, was employed [25].

Furthermore, it is to be discussed whether the differences between the second and third model for alkane transport into the cell are of such a significance or not. As far as it is known (cf. 2.2 or 2.1) there are not structural differences between attached and dissolved biosurfactants of one and the same microorganism. Alterations of the adsorption kinetics [196] indicate different mechanisms of alkane uptake by cells mediated by biosurfactants. This may be dependent on the physiological state of the cell, on its metabolism, on its appropriate phase of cell cycle and on the location of the biosurfactants as well. These facts implicate the speculation that the two models of alkane uptake can be considered to be the limits of one and the same general mechanism of alkane uptake in which biosurfactants or other alkane emulsifying factors are involved.

All the presented examples suggest that hydrocarbon solubilization by extracellular biosurfactants or membrane fragments could satisfactorily account for the substrate (alkane) uptake and growth.

Here we should reiterate that, as shown in Table 1, as a rule biosurfactants were not exclusively generated during growth on water-insoluble carbon sources but also on water-soluble substrates, such as glucose or glycerol. Furthermore, the largest yields of biosurfactants were obtained in the stationary growth phase [191, 192]. For most of the biosurfactants the synthesis is connected with or caused by an altered cell metabolism. This supposition may be valid and synthesis of biosurfactants should, therefore, be always independent of alkane substrate. Results of Cooper and Paddock [199, 200] with *Torulopsis bombicola* and *T. petrophilum* confirm this conclusion.

4.2 Biological Activities of Biosurfactants

Reports on biological activities of biosurfactants consider their biocidic activities. In this connection a review about amphiphiles of bacterial cell surfaces by Wicken and Knox [76] should be cited. Even if the matter in question does not directly deal with biosurfactants for the understanding of some of their effects, some aspects may be important. The authors pointed out that most consequences of amphiphile-eukaryotic cell interaction such as pyrogenicity and lethal toxicity, immunogenicity, mitogenicity, and other effects on other eukaryotic cellular systems, are closely related to the lipid moiety in the amphiphile (lipopolysaccharides, lipoteichoic acids). The binding of such compounds to erythrocytes, for instance, is based on lipid-lipid hydrophobic interactions. Such amphiphilic substances exhibit specific stimulatory or inhibitory effects on enzymes [76, 203, 204].

4.2.1 Mycolic Acid-Type and Similar Biosurfactants from *Actinomycetes*

The best known extracellular lipids seem to be the glycolipids of *Mycobacteria*, *Nocardia*, and *Corynebacteria*. This comprises the cord factor (trehalose 6,6′-dimycolate) and the mycolic acids being reviewed in Lederer's classical paper [205].

The highly toxic cord factor produced by *M. tuberculosis*, *Corynebacterium diphtheriae* and others has inhibitory effects on the migration of leucocytes [206, 207]. The activity of pyridine nucleotide-dependent dehydrogenases is decreased by the cord factor [208]. The glycolipids of *C. diphtheriae* cause mitochondrial damage. It influences the function of the mitochondrial membrane by inducing a decrease in respiration and a loss of oxidative phosphorylation [209] causing partly the pathogeneicity of *M. tuberculosis*. Additionally, glycomycolates may exhibit tumorostatic [210, 211] and immunogenic effects [212].

Much work was done with plant pathogenic *Corynebacteria*. The natural habitat of a few species of *Corynebacteria* is the surface of leaves. These bacteria which are predominantly host-specific [213] produce various metabolites including "toxins", polysaccharides, hormones and possibly enzymes [214–217]. Among the "toxins" biosurfactants may play an important role. These compounds may promote the adhesion to plant surfaces and possibly enhance an entry into the plant cell. But no correlation between biosurfactant production and pathogeneicity could be established [23]. The "annual ryegrass toxicity" accompanied with lesions in the central nervous system in rats and sheep is caused by a group of at least eight closely related glycolipids [218–220]. They are produced by *Corynebacterium rathayi* in cell cultures [221] as well as in vivo. Eight corynetoxins have been partially characterized; two have been crystallized. All are very similar in their physico-chemical and toxic properties. The major components appear to be of glycolipid character: both hydroxylated and unsaturated fatty acids (C_{17}) as well as hexosamines had been identified among the hydrolysis products. Therefore, *C. rathayi*-toxins differ from those described for the glycolipids of *C. diphtheriae* in their chemical structure. The minimal lethal dose for two-week-old rats was less than 0.2 mg kg^{-1} for six of seven fractions tested [219].

4.2.2 Biocidic Effects of Biosurfactants from Non-*Actinomycetes*

Vogel et al. [220] discussed a tunicamycin-like effect of these corynetoxins based on comparative investigations. In the 1960's, the term 'antibiotic' had already been connected to a product of *P. aeruginosa*, namely pyolipic acid [223] which is now known as rhamnolipid. Because of the nearly identical structure of the extracellular mannosylerythritol lipid from *Candida* sp. [224] and the acetylated mannosylerythritol antibiotics produced by *S. melanogramma* [63] and derivatives of the same compound which were found in *U. nuda* [225, 226], antibiotic activities also should be expected for the former compound. The biological activities of other biosurfactants justify this classification. In Table 5 some of the known effects of biosurfactants on bacterial growth are summarized. It is obvious that the different surfactants inhibit as a rule different taxa, but gram-positive bacteria such as bacilli, are most strongly effected whereas gram-negative bacteria are not so strongly or not influenced at all. This tolerance against surfactants may be due to differences in the molecular structure and composition of the outer membrane.

The antibiotic activity of the precursor of the di-rhamnosyl compound (R1), the mono-rhamnosyl (R2), is more substantial versus some gram-positive bacteria and *P. vulgaris*. Bergström et al. [227] detected the effect of rhamnolipid versus *M. tuberculosis*. Concentrations of this compound between 30 and 100 $\mu g\ ml^{-1}$ exhibited bactericidal effects on *M. tuberculosis*. Antibiotic effects have also been reported for biosurfactants from *Torulopsis* [51]. Furthermore, Ito et al. [127] showed mycoplasmacidal activities of R1 and R2 versus *Mycobacterium laidlawii* and *Mycobacterium gallisepticum* with concentrations between 250 $\mu g\ ml^{-1}$ and 10 $mg\ ml^{-1}$. The R1-fraction was most effective. These authors could also detect antiviral activities against vaccinia virus at low concentrations (20–50 $\mu g\ ml^{-1}$). The acute toxicities to mice were $LD_{50} > 400\ mg\ kg^{-1}$ of R1 and 117.5 $mg\ kg^{-1}$ of R2, respectively. In addition to these effects the *Pseudomonas* rhamnolipid exhibited hemolytic activities [233] which are also known for the exolipid from *S. sioyaensis* [232]. For rhamnolipid which may be an important virulence mechanism in different infections [201, 202] it was shown functioning as a detergent to solubilize phospholipids [222]. Recently Shryock et al. [233] have demonstrated that *Pseudomonas* rhamnolipid stimulates both leukocyte chemotaxis (directed migration) and chemokinesis (enhanced random migration) but it can also impair a chemotactic response to N-formyl-methionyl-leucyl-phenylalanine. The intracellular ornithine-containing lipid from *Bordetella pertussis* [234] strongly agglutinated type A and B human and rabbit erythrocytes. This activity changes to hemolysis after removal of hexadecanoic acid. Hemolytic activities of this lipid seem to be connected with the hydroxy fatty acid residues. All of the cited lipids contain such acids. According to the conclusions of Kimura and Otsuka [232] that the hemolytic activity of siolipin is related to the cytolytic action [234] other biosurfactants may also have effects on tissues culture cells, protoplasts of various bacteria, mitochondria, lysosomes and so on. Surfactin [235] as well as two peptidolipids produced by *Bacillus circulans* [94] promote the lysis of outer membrane of *P. aeruginosa* by altering the peptidoglucan, followed by the penetration of lytic enzymes into the peptidoglucan layer.

Whereas siolipin accelerated the fibrin clot formation, surfactin caused a strong elongation of this reaction by inhibition of conversion of fibrin monomer to poly-

Table 5. Antibiotic spectra of several biosurfactants

Biosurfactant (Maximal concentration tested)	Test organisms: No effect	Test organisms: Effect	Minimal inhibitory concentration (μg ml⁻¹)	
Rhamnolipids [228]	*Escherichia coli*	*Bacillus subtilis*	R1	R2
			208	5.2[a]
(R1 1700 μg ml⁻¹ R2 833 μg ml⁻¹)	*Proteus vulgaris*	*Pseudomonas aeruginosa*	—	5.2
	Shigella sonnei			
	P. aeruginosa (only for R1)			
	Salmonella typhosa			
	Klebsiella pneumoniae			
Rhamnolipid [227]		*Mycobacterium tuberculosis*	30–100	
Surfactin [231] (500 μg ml⁻¹)	*B. subtilis*	*Mycobacterium avium*	10	
	Staphylococcus aureus	*Mycobacterium phlei*	5	
	E. coli	*Mycobacterium smegmatis*	10	
	P. aeruginosa	*Mycobacterium 607 SM-F*	5	
	Serratia marcescens			
Siolipin [232]	*St. aureus*	*B. subtilis*[b]		
	E. coli			
	P. vulgaris			
	S. marcescens			
	M. tuberculosis			
Corynetoxins [220] (100 μg ml⁻¹)	*Agrobacterium radiobacter*	*Bacillus cereus*	25	
	Chromobacterium violaceum	*St. aureus*	25	
	Enterobacter aerogenes	*Corynebacterium michiganense* pv. *iranicum*	250	
	Erwinia (4 spec.)			
	E. coli	*Corynebacterium rathayi* (3 isolates)	3, 4, 16	
	Pseudomona acidovorans			

Table 5. (continued)

Biosurfactant (Maximal concentration tested)	Test organisms		Minimal inhibitory concentration ($\mu g\ ml^{-1}$)
	No effect	Effect	
	Pseudomonas fluorescens	*Corynebacterium tritici*	0.125
	Sporobolomyces tosens		
	Aspergillus nidulans		
	Penicillium verruculosum		
	Trichoderma harzianum		
	Sordaria fimicola		
	Pythium irregulare		
	Phytophtora cinnamomi		
	Phomopsis leptostromiformis		
	Saccharomyces cerevisiae		

[a] Calculated from minimal concentration of partial inhibition of growth;
[b] Inhibition only on synthetic medium, but inactive in presence of organic compounds;
[c] Determined using microtitre plates

mer [58]. Arima et al. [231] observed a remarkable anticholesteric effect of surfactin. In experiments with hypercholesteric rats they obtained a decrease in total cholesterol content of plasma and liver.

Some of these results reported should also be expected from other structurally analogue peptide lipids such as polymycine or bacillomycine L [98].

A consequence of the presented results seems to be that knowledge about the effects of biosurfactants on biological systems is rather limited. The biological activities may perhaps be compared with results obtained by studying extracellular lipids from *Cladosporium resinae* [112]. This organism excreted dodecanoic acid from which is known to inhibit biodegradation of petroleum by bacteria [236]. Additionally, fatty acids are inhibitors for certain bacteria [237,238] and yeasts [239] and can sensitize bacteria to the action of lytic enzymes [240] by physically dissolving a portion of the cells hydrophobic envelope. However, lipids may also certainly function as a metabolic effector — for example by the activation of autolysins. This fact has been recently demonstrated by the effects of dodecylglycerol on *Streptococcus faecium* ATCC 9790 [241]. However, the stimulation of the autolysin activity is attained by the activation of endogenous protease activity resulting in prevention of bacterial growth [242]. Therefore, extracellular fatty acids or lipids, in general, may gain ecological significance by inhibiting organisms that compete with the slowly growing *C. resinae* for example.

The antibacterial effects of biosurfactants, the generation of such lipids both on water-soluble and on water-insoluble carbon sources, and the inhibition of microbial growth on alkanes by those compounds allow the conclusion that biosurfactants might also possess such a biological function as described above. The excretion of biosurfactants may be one response to altered environmental conditions which may be expressed, e.g., by limitations (cf. 2.1) to survive and compete successfully with other microorganisms.

5 Applications and Concluding Remarks

As shown above, microorganisms produce a variety of surface active metabolites which are often potent emulsifiers. For this reason, most studies have been performed under the aspect of applications [1,16,101,115,116,118,228,245,246,256]. This is done to meet the demands for surfactants for industrial and domestic use. The potential applications and most probable functions are summarized in Table 6.

The respective properties depend on the appropriate physico-chemical properties of each compound. This means, that an application in one of the fields shown always demands special biosurfactants which exhibit the appropriate properties. There is, however, a lack of appropriate methods to determine these respective properties. The most common used approximation is the distinction of surfactants based on their HLB-values (cf. 2.2). According to this classification, structurally appropriate emulsifying properties may be derived. The HLB-system like other comparable systems that aim at a connection between the structure of surfactants with the properties to be expected on the basis of the known chemical structure, i.e. emulsification of apolare phases in water or vice versa, is not sufficient to obtain satisfactory information concerning technical applicative properties of the lipid in question. Properties of the resulting emulsions such as long-term stability, pH stability, temperature stability and so

Table 6. Potential commercial applications of biosurfactants (adapted from Kosaric et al. [77])

	Emusification, de-emulsification	Wetting, spreading, penetration	Solubilization, solids dispersal	Air entrainment, foaming	Defoaming	Detergency	Antistatic	Corrosion inhibition	Antibiotic
Agriculture	×	×	×						×
Biotechnology	×	×	×			×		×	×
Building and construction		×		×					
Cosmetics, pharmaceutics	×	×		×					×
Food and beverage	×	×	×	×	×	×			
Industrial cleaning		×				×		×	×
Leather and textiles	×	×	×			×	×		
Mining		×	×	×				×	
Paper		×			×	×			
Paint and protective coatings	×	×					×		
Petroleum and petrochemical industry	×	×	×	×	×	×	×	×	
Transport (oil, coal by pipelines)	×	×	×					×	
Environmental save	×		×	×	×				
Dewaterization of coal	×			×					
Protecting against fire		×		×					

forth may only be taken on the basis of experimental work. It is quite reasonable too, that the potential application fields will be restricted by properties of the surfactant in question. A disadvantage of biosurfactants that seems to restrict their application represents in most cases the genetically determined structure of the biosurfactant produced by one microorganism. A variation of the general composition of these compounds and of the properties is only possible in a few cases (cf. 2.2). This disadvantage may be overcome by chemical modifications which should offer virtually unlimited properties. It must be added that low concentrations and long term production processes are typical features for reports dealing with microbial biosurfactant production.

The advantages of biosurfactants are the following:

— These compounds can be produced under application of relatively simple and inexpensive procedures.

— New structural types of surfactants can be obtained which are not easily synthesized by chemical processes [1].

— A positive ecological aspect, due to complete biodegradability.

— For specific application, specific properties are accumulated (for instance the micelle-forming ability connected with antibiotic effects (cf. 4.2)).

The applications presented in Table 6 are in most cases potential uses. However it should be noted that wherever emulsions are to be formed or broken, where processes occur at interfaces and/or are facilitated by interfaces, the application of biosurfactants is possible.

Nowadays, the enhanced oil recovery (EOR) or the petroleum industry as a whole seem to be the most important field, in which biosurfactants are applicable. The technical production and the application of biosurfactants for EOR has been repeatedly reported in patents. Wagner et al. [25,121,247] obtained an increase of 30% of the total oil yield by supplementation of trehalose lipids (50 mg l^{-1}, related to the trehalose content) from *Nocardia rhodochrous*. Other examples are described by Gerson and Zajic [1].

The application of biosurfactants is limited by appropriate demands: For EOR-formulations which generate low γ_i (10^{-3}–10^{-5} mN m^{-1}) in situ (formation of micro-emulsions) [244] and which exhibit phase inversion temperatures that are specifically for the temperature of the oil deposits are preferred (cf. [53]). These demands may be met by manipulation of physiological and genetic properties of the microorganisms which provide the desired properties of the biosurfactants.

Finnerty and Singer [249] stressed that useful biosurfactants may be involved in a number of recovery, production and transport problems in the oil industry. Furthermore, other aspects like the complementation of synthetic surfactants by biosurfactants for EOR and the combination with co-surfactants have to be considered.

Other applications are connected with the properties discussed in Chapts. 2 and 4. Analogous to the use of synthetic surfactants in the field of wetting of coal [250,251] for instance, in principal, all of the areas known for those are potential fields of biosurfactants use. This may also cover their role in alkane utilization and the other biological effects for biotechnological industry, such as the cultivation of cells on liquid interfaces as shown for fibroblasts in connection with pentafluorobenzoyl chloride as surfactant [252] or as additiva in pharmaceutical products: The peptide lipid surfactin (cf. Table 1) does not inhibit the fungus *Trichophyton mentygrophytes* which is responsible for dermatophytose disease [231]. In combination with pyrrolnitrine, surfactin increased the antifungal activity and protected the inactivation of this fungicide in serum. In other situations surface-active substances play an important role in the protection against *Schistosoma cercaria* [243] or *Trypanosoma* sp. [248].

Furthermore, there are many other potential and possible applications as shown in Table 6. However, as a rule, two areas of interest should be satisfied: The already discussed commercial and the academic one. The latter area is not well developed up to now. Most work comprises the description of phenomenological effects. Unfortunately, comparatively little has been published on the metabolic pathways, their principles of regulation on epigenetic or genetic levels, or on the manipulation of the appropriate metabolic steps. To meet these demands (including the commercial ones) it is important to investigate biosynthetic principles, appropriate enzymes, their regulation and the connection with lipid metabolism as a whole. As discussed by Finnerty and Singer [249] with regard to the enhanced oil recovery, it is also valid that "the future interface between . . . engineering and the application of biological processes is complementary in terms of contributions that the biologist can offer the industry". The technical problems, such as production or isolation will be solved by further investigations. Applications will become valuable enough when the processes of biosurfactant production is economical and production rates meet the commercial demands. Therefore, the biotechnology of biosurfactants seems to be a challenge for the development of future technology today.

6 Nomenclature and Definitions

AMP	adenosine-5′-monophosphate
CoA	coenzyme A
NAD^+	nicotinamide adenine dinucleotide (oxidized)
TDP	thymidine-5′-diphosphate
UDP	uridine-5′-diphosphate
CMC	critical micelle concentration
C:N	carbon-nitrogen ratio in growth medium
C:P	carbon-phosphate ratio in growth medium
EOR	enhanced oil recovery
HLB	hydrophile-lipophile balance
PIT	phase inversion temperature
o/w	oil-in-water emulsion
w/o	water-in-oil emulsion
γ_i	interfacial tension
γ_s	surface tension

7 References

1. Gerson, D. F., Zajic, J. E.: Process Biochem. *10*, 20 (1979)
2. Layman, P. L.: Chem. Eng. News *60*, 13 (1982)
3. Diekmann, H.: Nukleotid-Fermentationen mit *Brevibacterium ammoniagenes*, in: Fermentation (ed. Lafferty, R. M.) p. 121, Wien: Springer Verlag 1981
4. Bartha, R., Atlas, R. M.: Adv. Appl. Microbiol. *22*, 225 (1977)
5. Cooney, J. J., Summers, R. J.: Hydrocarbon-using microorganisms in three freshwater ecosystems, in: Proc. of the IIIrd Intern. Biodegr. Symp. (eds. Sharpley, J. M., Kaplan, A. M.) p. 141, London: Applied Sciences Publishers 1976
6. Austin, B. et al.: Appl. Environ. Microbiol. *34*, 60 (1977)
7. Polyakova, J. M.: Mikrobiologija *31*, 1076 (1962)
8. Atlas, R. M. et al.: in: Proceedings of the Symp. on Assessment of the Environ. Impact of Accidental Oil Spills in the Oceans. Ann Arbor: Science Publ. 1980
9. Bhosle, N. B., Mavinkurve, S.: Marine Environ. Res. *4*, 53 (1980)
10. Platpira, W. P.: Latvijas PSR Zinatum Akademijas Vestis *3*, 80 (1982)
11. Jarvis, F. G., Johnson, M. J.: J. Am. Chem. Soc. *71*, 4124 (1949)
12. Tulloch, A. P., Spencer, J. F. T.: Can. J. Chem. *42*, 830 (1964)
13. Tulloch, A. P. et al.: ibid. *40*, 1326 (1960)
14. Tulloch, A. P., Spencer, J. F. T.: ibid. *46*, 1523 (1968)
15. Gorin, P. A. J., Spencer, J. F. T., Tulloch, A. P.: ibid. *39*, 846 (1961)
16. Cooper, D. G., Zajic, J. E.: Adv. Appl. Microbiol. *26*, 229 (1980)
17. Suzuki, T., Tanaka, H., Itoh, S.: Agric. Biol. Chem. *38*, 557 (1974)
18. Itoh, S., Suzuki, T.: ibid. *38*, 1443 (1974)
19. Gerson, D. F., Zajic, J. E.: Antonie van Leeuwenhoek *45*, 81 (1979)
20. Macdonald, C. R., Cooper, D. G., Zajic, J. E.: Appl. Environ. Microbiol. *41*, 117 (1981)
21. Duvnjak, Z. et al.: Biotech. Bioeng. *24*, 165 (1982)
22. Margaritis, A., Zajic, J. E., Gerson, D. F.: ibid. *21*, 1151 (1979)
23. Akit, J. et al.: Current Microbiol. *6*, 145 (1981)
24. Cooper, D. G., Zajic, J. E., Gerson, D. F.: Appl. Environ. Microbiol. *37*, 4 (1979)
25. Wagner, F., Bock, H., Kretschmer, A.: Gewinnung von Tensiden mit n-Alkan-oxidierenden Mikroorganismen, in: Fermentation (ed. Lafferty, R. M.) p. 181, Wien: Springer Verlag 1981
26. Lederer, E.: Angew. Chem. *72*, 372 (1960)

27. Asselineau, J.: The Bacterial Lipids, San Francisco: Holden-Day 1966
28. Cundell, A. M.: Develop. Ind. Microbiol. *15*, 250 (1974)
29. Mulkins-Philipps, G. J. et al.: Can. J. Microbiol. *20*, 955 (1974)
30. Pareilleux, A.: Eur. J. Appl. Microbiol. Biotechnol. *8*, 91 (1979)
31. Illarionova, V. J. et al.: Mikrobiologija *53*, 423 (1984)
32. Käppeli, O.: Ph. D. Thesis, Diss. No. 5661 E.T.H. Zürich (1976)
33. Reddy, P. G. et al.: Biotech. Bioeng. *25*, 387 (1983)
34. Itoh, S., Suzuki, T.: Agric. Biol. Chem. *36*, 2233 (1972)
35. Hisatsuka, K. et al.: ibid. *35*, 684 (1971)
36. Hisatsuka, K. et al.: ibid. *35*, 686 (1971)
37. Hisatsuka, K. et al.: ibid. *36*, 1361 (1972)
38. Hisatsuka, K. et al.: ibid. *41*, 445 (1977)
39. Yamaguchi, M.: Chem. Ind. *17*, 741 (1976)
40. Burger, M. M., Glaser, L., Burton, R. M.: J. Biol. Chem. *238*, 2595 (1963)
41. Hauser, G. et al.: Fed. Proc. *12*, 215 (1953)
42. Hauser, G. et al.: J. Bacteriol. *68*, 645 (1954)
43. Guerra-Santos, L., Käppeli, O., Fiechter, A.: Appl. Environ. Microbiol. *48*, 301 (1984)
44. Haferburg, D. et al.: DD-WP 269 261 2, 08. 11. 1984
45. Syldatk, C. et al.: Z. Naturforsch. *40c*, 61 (1985)
46. Li, Z. Y. et al.: Appl. Environ. Microbiol. *48*, 610 (1984)
47. Zajic, J. E., Guignard, H., Gerson, D. F.: Biotech. Bioeng. *19*, 1285 (1977)
48. Zajic, J. E., Guignard, H., Gerson, D. F.: ibid. *19*, 1303 (1977)
49. Haurowitz, F.: Fed. Proc. *12*, 215 (1953)
50. Tulloch, A. P. et al.: Can. J. Chem. *46*, 3337 (1968)
51. Hommel, R. et al.: DD-WP 269 259 8, 08. 11. 1984
52. Botham, P. A., Ratledge, C.: J. Gen. Microbiol. *114*, 361 (1979)
53. Tulloch, A. P., Spencer, J. F. T., Deinema, M. H.: Can. J. Chem. *46*, 345 (1968)
54. Wickerham, L. J.: J. Bacteriol. *80*, 484 (1960)
55. Stodola, F. H., Wickerham, L. J.: J. Biol. Chem. *235*, 2584 (1960)
56. Takahara, Y. et al.: Agric. Biol. Chem. *40*, 1901 (1976)
57. Cooper, D. G. et al.: Appl. Environ. Microbiol. *42*, 408 (1981)
58. Arima, K., Kakinuma, A., Tamura, G.: Biochem. Biophys. Res. Commun. *31*, 488 (1968)
59. Kakinuma, A. et al.: Agric. Biol. Chem. *33*, 971 (1969)
60. Kakinuma, A. et al.: ibid. *33*, 973 (1969)
61. Weinberg, E. D.: Adv. Microbial Physiology *4*, 1 (1970)
62. Kawashima, H. et al.: J. Ferment. Technol. *61*, 143 (1983)
63. Deml, G. et al.: Phytochemistry *19*, 83 (1980)
64. Okazaki, H. et al.: Agric. Biol. Chem. *33*, 764 (1969)
65. La Riviere, J. W. M.: Antonie van Leeuwenhoek *21*, 1 (1955)
66. Knoche, H. W., Shiveley, J. M.: J. Biol. Chem. *247*, 170 (1972)
67. Tahara, Y. et al.: Agric. Biol. Chem. *40*, 243 (1976)
68. Kawanami, J. et al.: Chem. Phys. Lipids *7*, 159 (1971)
69. Käppeli, O., Finnerty, W. R.: J. Bacteriol. *140*, 707 (1979)
70. Käppeli, O., Finnerty, W. R.: Biotech. Bioeng. *22*, 495 (1980)
71. Biotechnology A.G.: JP-P. 80, 112, 201; 29. 08. 1980
72. Zuckerberg, A. et al.: Appl. Environ. Microbiol. *37*, 414 (1979)
73. Rosenberg, E. et al.: ibid. *37*, 409 (1979)
74. Rosenberg, E. et al.: ibid. *37*, 402 (1979)
75. Kessler, R. E.: J. Bacteriol. *137*, 869 (1979)
76. Wicken, A. J., Knox, K. W.: Biochim. Biophys. Acta *604*, 1 (1980)
77. Kosaric, N., Gray, N. C. C., Cairns, W. L.: Microbial emulsifiers and de-emulsifiers, in: Biotechnology Vol. 3, p. 575, Weinheim: Verlag Chemie 1983
78. Tanaka, A. et al.: J. Ferment. Technol. *46*, 468 (1968)
79. Edwards, J. R., Hayashi, J. A.: Arch. Biochem. Biophys. *111*, 415 (1965)
80. Hirayama, T., Kato, J.: FEBS Lett. *139*, 81 (1982)
81. Jones, D. F.: J. Chem. Soc. (C) 479 (1967)
82. Jones, D. F., Howe, R.: ibid. 2801 (1968)

83. Jones, D. F.: J. Chem. Soc. 2809 (1968)
84. Jones, D. F., Howe, R.: ibid. 2816 (1968)
85. Jones, D. F., Howe, R.: ibid. 2827 (1968)
86. Tulloch, A. P., Spencer, J. F. T., Deinema, M. H.: Can. J. Chem. *46*, 345 (1968)
87. Frautz, B., Lang, S., Wagner, F.: Eur. Cong. Biotech. 3rd, Abstracts I-79 (1984)
88. Anderson, R. J., Newman, M.: J. Biol. Chem. *101*, 499 (1933)
89. Wilkinson, S. G.: Biochim. Biophys. Acta *270*, 1 (1972)
90. Tahara, Y., Yamada, Y., Kondo, K.: Agric. Biol. Chem. *40*, 1449 (1976)
91. Tahara, Y., Yamada, Y., Kondo, K.: ibid. *39*, 2261 (1975)
92. Beebe, J. L., Umbreit, W.: J. Bacteriol. *108*, 612 (1971)
93. Thiele, O. W., Oulevey, J., Hunneman, D. H.: Eur. J. Biochem. *139*, 131 (1984)
94. Bernheimer, A. W., Avigad, L.: J. Gen. Microbiol. *61*, 361 (1970)
95. Tahara, J. et al.: Agric. Biol. Chem. *40*, 243 (1976)
96. Godchaux, W., Leadbetter, E. R.: J. Biol. Chem. *259*, 2983 (1984)
97. Pepoux, T., Michel, G., Decambe, L.: Eur. J. Biochem. *63*, 391 (1976)
98. Besson, F. et al.: ibid. *77*, 61 (1977)
99. Walton, R. B. et al.: J. Clin. Invest. *28*, 924 (1949)
100. Heinemann, B. et al.: Antibiot. Chemother. *3*, 1239 (1953)
101. Cooper, D. G.: J. Bacteriol. *137*, 795 (1979)
102. Kretschmer, A., Wagner, F.: Biochim. Biophys. Acta *753*, 306 (1983)
103. Lindam, E. et al.: J. Biochem. Microbiol. Technol. Eng. *3*, 219 (1961)
104. Odier, E.: Ann. Microbiol. *127b*, 213 (1976)
105. Makula, R. A., Finnerty, W. R.: J. Bacteriol. *112*, 398 (1972)
106. Shiio, J. et al.: Agric. Biol. Chem. *35*, 2033 (1971)
107. Izuka, H. et al.: J. Gen. Appl. Microbiol. *12*, 131 (1966)
108. Uchio, R. et al.: Agric. Biol. Chem. *36*, 426 (1972)
109. Abbot, B. J. et al.: Adv. Appl. Microbiol. *14*, 249 (1971)
110. Heringa, J. W. et al.: Antonie van Leeuwenhoek *27*, 51 (1981)
111. Kester, A. S. et al.: J. Bacteriol. *85*, 859 (1963)
112. Siporin, C., Cooney, J. J.: Appl. Microbiol. *29*, 604 (1975)
113. Shively, J. M. et al.: J. Bacteriol. *94*, 1679 (1967)
114. Nakao, Y. et al.: Agric. Biol. Chem. *37*, 2399 (1973)
115. Kikuchi, M., Kanamura, T., Nakao, Y.: ibid. *37*, 2405 (1973)
116. Cooper, D. G., Zajic, J. E., Denis, C.: J. Am. Oil Chem. Soc. *58*, 77 (1981)
117. Ristau, E., Wagner, F.: Biotechnol. Lett. *5*, 95 (1983)
118. Gerson, D. F., Zajic, J. E.: Dev. Ind. Microbiol. *19*, 577 (1978)
119. Zajic, J. E., Knetting, E.: US-P. 3,997,398; 14. 12. 1976
120. Tanaka, K., Suzuki, T.: US-P. 3,637,461; 25. 01. 1972
121. Wagner, F. et al.: US-P. 4,286,660; 01. 09. 1981
122. Zajic, J. E. et al.: US-P. 4,355,109; 19. 10. 1982
123. Spencer, J. F. T. et al.: US-P. 3,445,337; 20. 05. 1969
124. Spencer, J. F. T. et al.: US-P. 3,312,684; 04. 04. 1967
125. Tanaka, K., Suzuki, T.: DE-AS 1,965,975; 31. 01. 1969
126. Tanaka, K., Suzuki, T.: DE-AS 1,905,472; 31. 01. 1969
127. Ito, S. et al.: J. Antibiot. *24*, 855 (1971)
128. Esders, T. W., Light, R. J.: J. Biol. Chem. *247*, 7494 (1972)
129. Esders, T. W., Light, R. J.: ibid. *247*, 1375 (1972)
130. Goldman, D. S., Lornitzo, F. A.: ibid. *237*, 3332 (1962)
131. Elbein, A. D.: Adv. Carbohydr. Chem. Biochem. *30*, 227 (1974)
132. Loomis, S. H., Madin, K. A. C., Crowe, J. H.: J. Exp. Zoology *211*, 311 (1980)
133. Walker, R. W., Prome, J. C., Lacave, C. S.: Biochim. Biophys. Acta *326*, 52 (1973)
134. Kretschmer, A.: Ph.D. Thesis, Techn. Univ. Braunschweig (1981)
135. Ahibo-Coffy, A. et al.: Chem. Phys. Lipids *22*, 185 (1978)
136. Brennan, P. J., Lehane, D. P., Thomas, D. W.: Eur. J. Biochem. *13*, 117 (1970)
137. Nakahara, T., Kawashima, H., Tabuchi, T.: in: Advances in Biotechnology (eds. Moo-Young, M., Robinson, C. W., Vecina, C.) Vol. 1, pp. 149–153, Toronto: Pergamon Press 1981
138. Promè, J. C., Walker, R. W., Lacave, C.: Comptes Rendus Acad. Sci. Serie C t. 278, 1065 (1974)

139. Fulco, J. A.: Prog. Lipid Res. *22*, 133 (1983)
140. Dworkin, M., Keller, K. H., Weisberg, D.: J. Bacteriol. *155*, 1367 (1983)
141. Rosenberg, M.: FEMS Microbiol. Lett. *22*, 289 (1984)
142. Egge, H.: in: Etherlipids: Biochemical and Biomedical Aspects, (eds. Mangold, H. K., Paltauf, F.) pp. 141–159, New York: Academic Press 1983
143. Cory, V. F. et al.: Biotech. Bioeng. *13*, 825 (1971)
144. Shennan, J. L., Levi, J. D.: Prog. Ind. Microbiol. *13*, 1 (1974)
145. Watkinson, R. J.: in: Hydrocarbons in Biotechnology (eds. Harrison, D. E., Higgins, I. J., Watkinson, R. J.) pp. 11–24, London: Heyden and Son 1980
146. Ratledge, C.: in: Developments in Biodegradation of Hydrocarbons (ed. Watkinson, R. J.) pp. 1–46, London: Appl. Sci. Publ. 1978
147. Levi, J. D., Shennan, J. L., Ebbon, G. P.: Econ. Microbiol. *4*, 361 (1979)
148. Miura, Y.: Adv. Biochem. Eng. *9*, 31 (1978)
149. Goma, G., Pareilleux, A., Durand, G.: J. Ferment. Technol. *51*, 616 (1976)
150. Roy, P. K. et al.: Biotech. Bioeng. *21*, 955 (1979)
151. Mallee, F. M. et al.: ibid. *19*, 1793 (1977)
152. Nakahara, T. et al.: ibid. *19*, 9 (1977)
153. Einsele, A. et al.: Adv. Biochem. Eng. *1*, 169 (1971)
154. Kulkarni, K. et al.: Biotech. Bioeng. *21*, 337 (1979)
155. Einsele, A., Blanch, H. D., Fiechter, A.: ibid. Symp. *4*, 455 (1973)
156. Milira, Y. et al.: ibid. *19*, 701 (1977)
157. McLee, A. G., Davies, S. L.: Can. J. Microbiol. *18*, 315 (1972)
158. Käppeli, O., Fiechter, A.: Biotech. Bioeng. *18*, 967 (1974)
159. Kennedy, R. S. et al.: Arch. Microbiol. *102*, 75 (1975)
160. Rosenberg, M., Gutnick, D., Rosenberg, E.: FEMS Microbiol. Lett. *9*, 29 (1980)
161. Stuber, W.: Ph.D. Thesis, Karl-Marx-Universität Leipzig (1982)
162. Kefford, B., Kjelleberg, S., Marshall, K.: Arch. Microbiol. *133*, 257 (1982)
163. Miura, Y. et al.: Biotech. Bioeng. *19*, 715 (1977)
164. Mimura, A., Watanabe, S., Takeda, I.: J. Ferment. Technol. *49*, 255 (1971)
165. Rosenberg, M., Rosenberg, E., Gutnick, D.: in: Microbial Adhesion to Surfaces (eds. Berkeley, R. C. W., Lynch, J. M., Helling, J., Rutter, P. R., Vincent, E.) Publ. for the Soc. of Chem. Ind. pp. 541–542, London 1980
166. Käppeli, O., Müller, M., Fiechter, A.: J. Bacteriol. *133*, 952 (1978)
167. Käppeli, O., Fiechter, A.: ibid. *131*, 917 (1977)
168. Rosenberg, M. et al.: Appl. Environ. Microbiol. *44*, 929 (1982)
169. Scott, C. C. L., Finnerty, W. R.: J. Bacteriol. *127*, 481 (1976)
170. Claus, R., Käppeli, O., Fiechter, A.: J. Gen. Microbiol. *130*, 1035 (1984)
171. Tiurin, V. S., Ilchenko, A. P., Dimitriew, V. V.: Mikrobiologija *44*, 481 (1975)
172. Petriuevic, S. B., Dovgum, L. I.: ibid. *49*, 78 (1980)
173. Aiba, S. et al.: J. Ferment. Technol. *47*, 211 (1969)
174. Müller, H. et al.: Z. Allg. Mikrobiol. *23*, 645 (1983)
175. Moo-Young, M., Shimizu, T., Whitworth, D. A.: Biotech. Bioeng. *13*, 841 (1971)
176. Erickson, L. E., Nakahara, T.: Proc. Biochem. *10*, 9 (1975)
177. Gutnick, D. L., Rosenberg, E.: Annu. Rev. Microbiol. *31*, 379 (1977)
178. Koronelli, T. V.: Acta Biotechnol. *3*, 309 (1983)
179. Zajic, J. E., Panchal, C. J.: CRC Crit. Rev. Microbiol. *5*, 39 (1976)
180. Tanaka, A., Fukui, S.: J. Ferment. Technol. *49*, 809 (1971)
181. Wang, D. I. C., Ochoa, A.: Biotech. Bioeng. *14*, 345 (1972)
182. Whitworth, D. A., Moo-Young, M., Viswanatha, T.: ibid. *15*, 649 (1973)
183. Chakravarty, M. et al.: ibid. *17*, 399 (1975)
184. Whitworth, D. A.: Compt. Rend. Trav. Lab. Carlsberg *40*, 33 (1974)
185. Nakahara, T., Hisatsuka, K., Minoda, V.: J. Ferment. Technol. *59*, 415 (1981)
186. Nakahara, T. et al.: ibid. *61*, 19 (1983)
187. Hisatsuka, K. et al.: ibid. *39*, 999 (1975)
188. Koronelli, T. V., Koharova, T. I., Ignatshenko, A. V.: Mikrobiologija *52*, 94 (1983)
189. Banerjee, S., Duttagupta, S., Chakrabarty, A. M.: Arch. Microbiol. *135*, 110 (1983)
190. Inoue, S., Ito, S.: Biotechnol. Lett. *4*, 3 (1982)

191. Ito, S., Inoue, S.: Appl. Environ. Microbiol. *43*, 1278 (1982)
192. Ito, S., Kinta, M., Inoue, S.: Agric. Biol. Chem. *44*, 2221 (1980)
193. Iguchi, T., Takeda, I., Ohsaiwa, H.: ibid. *33*, 1657 (1969)
194. Bayer, E. A., Rosenberg, E., Gutnick, D. L.: J. Gen. Microbiol. *127*, 295 (1981)
195. Reddy, P. G. et al.: Biotech. Bioeng. *25*, 387 (1983)
196. Neufeld, R. J., Zajic, J. E., Gerson, D. F.: Appl. Environ. Microbiol. *39*, 511 (1980)
197. Rapp, P. et al.: J. Gen. Microbiol. *115*, 491 (1979)
198. Kretschmer, A., Bock, H., Wagner, F.: Appl. Environ. Microbiol. *44*, 864 (1982)
199. Cooper, D. G., Paddock, D. A.: ibid. *46*, 1426 (1983)
200. Cooper, D. G., Paddock, D. A.: ibid. *47*, 173 (1984)
201. Al-Dujaili, A. H.: J. Hygiene *77*, 211 (1976)
202. Johnson, M. K., Allen, J. H.: Invest. Ophthalmal. Visual Sci. *17*, 480 (1978)
203. Schachtele, C. F. et al.: Infect. Immun. *22*, 714 (1978)
204. Romeo, D., Girard, A., Rothfield, L.: J. Mol. Biol. *53*, 475 (1970)
205. Lederer, E.: Chem. Phys. Lipids *1*, 294 (1967)
206. Bloch, H.: J. Exp. Med. *91*, 197 (1950)
207. Bloch, H.: ibid. *92*, 507 (1950)
208. Kato, M. et al.: Am. Rev. Respirat. Diseases *80*, 240 (1955)
209. Kato, M.: Arch. Biochem. Biophys. *140*, 379 (1970)
210. Meyer, T. J. et al.: J. Natl. Cancer Inst. *52*, 103 (1974)
211. Ribi, E. et al.: ibid. *55*, 1253 (1975)
212. Jolles, P., Samouz, D., Lederer, E.: Arch. Biochem. Biophys. Suppl. *1*, 283 (1962)
213. Diaz-Maurino, T., Perkins, H. R.: J. Gen. Microbiol. *80*, 533 (1974)
214. Bateman, D. R., Basham, H. G.: in: Physiological Plant Pathology (eds. Heitefuss, R., Williams, P. H.) pp. 316–355, New York: Springer Verlag 1976
215. Patil, S. S.: Ann. Rev. Phytopathol. *12*, 259 (1974)
216. Strobel, G. A.: Ann. Rev. Microbiol. *31*, 205 (1977)
217. Vidaver, A. K., Statt, M. P.: in: The Prokaryotes; A Handbook on Habitat, Isolation, and Identification of Bacteria (eds. Starr, M. P. et al.) Vol. 2, pp. 1878–1887, Berlin: Springer Verlag 1981
218. Vogel, P. et al.: Aust. J. Exp. Biol. Med. Sci. *59*, 455 (1981)
219. Berry, P. H., Vogel, P.: ibid. *60*, 129 (1982)
220. Vogel, P. et al.: Biochem. Biophys. Res. Commun. *105*, 835 (1982)
221. Stynes, B. A., Petterson, D. S.: Physiol. Plant. Path. *16*, 163 (1980)
222. Kurioka, S., Liu, P. V.: J. Bacteriol. *93*, 670 (1967)
223. O'Leary, W.: Bacteriol. Rev. *26*, 421 (1962)
224. Kawashima, H. et al.: J. Ferment. Technol. *61*, 143 (1983)
225. Haskins, R. H., Thorn, J. A., Boothroyd, D.: Can. J. Microbiol. *1*, 749 (1955)
226. Bhattacharjee, S. S., Haskins, R. H., Gorin, P. A. J.: Carbohyd. Res. *13*, 235 (1970)
227. Bergström, S., Theorell, H., Davide, H.: Arch. Biochem. *10*, 165 (1946)
228. Kosaric, N. et al.: J. Am. Oil Chem. Soc. *61*, 1735 (1984)
229. McAuliffe, C.: Science *163*, 478 (1969)
230. Franks, F.: Nature *210*, 87 (1966)
231. Arima, K., Tamura, G., Kasinuma, A.: DE-OS 1803 987, 19. 10. 1968
232. Kimura, A., Otsuka, H.: Agric. Biol. Chem. *33*, 1291 (1969)
233. Shryock, T. R. et al.: Current Microbiol. *10*, 323 (1984)
234. Bernheimer, A. W.: Science *159*, 847 (1968)
235. Kakinuma, A., Arima, K.: Kagabu to Seibutu *6*, 606 (1968)
236. Atlas, R. M., Bartha, R.: Antonie van Leeuwenhoek, J. Microbiol. Serol. *39*, 257 (1973)
237. Galbraith, H. et al.: J. Appl. Bacteriol. *34*, 803 (1971)
238. Nieman, C.: Bacteriol. Rev. *18*, 147 (1954)
239. Niwano, M. et al.: Agric. Biol. Chem. *48*, 1359 (1984)
240. Haska, G., Noren, B., Odham, G.: Physiol. Plant *27*, 181 (1972)
241. Ved, H. S. et al.: J. Biol. Chem. *259*, 8115 (1984)
242. Ved, H. S., Gustow, E., Pieringer, R. A.: ibid. *259*, 8122 (1984)
243. Combes, C., Arnaudis, J.: Acta Tropica *39*, 79 (1982)
244. Taber, J. J.: Pure Appl. Chem. *52*, 1323 (1980)

245. Guerra-Santos, L., Käppeli, O., Fiechter, A.: Proceedings Intern. Conf. Microbial Enhancement Oil Recovery pp. 12–14, Afton 1982
246. Singer, M. E. et al.: Proceedings Intern. Conf. Microbial Enhancement Oil Recovery pp. 94–101, Afton 1982
247. Wagner, F. et al.: DE-PS 30 04 003; 4. 2. 1980
248. Zumbühl, O., Weder, H. G.: Biochem. Biophys. Res. Commun. *107*, 869 (1982)
249. Finnerty, W. R., Singer, M. E.: Biotechnol. *1*, 47 (1983)
250. Esuhi, K., Meguro, K., Honda, H.: Bull. Chem. Soc. Jpn. *6*, 933 (1983)
251. Esuhi, K., Meguro, K., Honda, H.: ibid. *6*, 3169 (1983)
252. Keese, C. R., Giaever, I.: Proc. Natl. Acad. Sci. *80*, 5622 (1983)
253. Cirigliano, M. C., Carman, G. M.: Appl. Environm. Microbiol. *48*, 747 (1984)
254. Boulton, C. A., Ratledge, C.: in: Topics in Enzyme and Fermentation Biotechnology (ed. Wiseman, A.) Vol. 9, pp. 11–77, New York, Chichester, Brisbane, Toronto: John Wiley & Sons 1984
255. Bühler, M., Schindler, J.: in: Biotechnology (eds. Rehm, H.-J., Reed, G.) Vol. 6a, pp. 329–385, Weinheim: Verlag Chemie 1984
256. Zajic, J. E., Seffens, W.: CRC Crit. Rev. Biotechnol. *1*, 87 (1984)
257. Xiuyan, W. et al.: Acta Microbiologica Sinica *22*, 269 (1982)

Applications of Stereoselectivity of Enzymes: Synthesis of Optically Active Amino Acids and α-Hydroxy Acids, and Stereospecific Isotope-Labeling of Amino Acids, Amines and Coenzymes

Kazuo Yonaha
Department of Agricultural Chemistry, University of The Ryukyus Nishihara, Okinawa 903-01, Japan
Kenji Soda
Institute for Chemical Research, Kyoto University, Uji, Kyoto 611, Japan

1 Introduction ... 96
2 Enzymatic Synthesis of L-Amino Acids and α-Hydroxy Acids from Optically Active Starting Materials ... 98
2.1 Synthesis of *S*-Substituted L-Homocysteines with L-Methionine γ-Lyase ... 98
2.2 Synthesis of L-Tyrosine and its Related Compounds with β-Tyrosinase ... 102
2.3 Synthesis of L-Cysteine-related Amino Acids and Their Selenium Analogues with Tryptophan Synthase ... 104
2.4 Synthesis of Amino Acids and Their Derivatives with Other PLP-Enzymes ... 107
2.5 Synthesis of D-2-Hydroxy Acids with L-2-Haloacid Dehalogenase ... 108
3 Enzymatic Synthesis of Optically Active Amino Acids and α-Hydroxy Acids from Racemic Starting Materials ... 109
3.1 Synthesis of L- and D-Amino Acids by Asymmetric Derivatization of Racemates ... 109
3.2 Synthesis of L- and D-Amino Acids by Stereospecific Hydrolysis of Racemates ... 110
3.2.1 Production of Amino Acids from *N*-Acyl-DL-Amino Acids with Amino Acylase ... 110
3.2.2 Production of Amino Acids from DL-Hydantoins with Hydantoinase ... 111
3.2.3 Production of Amino Acids from Amino Acid Esters with Esterases ... 113
3.2.4 Production of Amino Acids from Amino Acid Amides with Amidases ... 114
3.3 Synthesis of Amino Acids and α-Hydroxy Acids by Stereospecific Resolution of Racemates by a Combination of Multiple Enzymes ... 114
3.3.1 Production of D-α-Aminobutyrate and D-Methionine from DL-Methionine with Methionine γ-Lyase and D-Amino Acid Transaminase ... 114
3.3.2 Production of L-Tryptophan from Indole and DL-Serine with Tryptophan Synthase and Amino Acid Racemase ... 115
3.3.3 Production of L-Lysine from DL-2-Amino-6-Caprolactam with L-2-Amino-6-caprolactamase and 2-Amino-6-caprolactam Racemase ... 115
3.3.4 Production of L-Cysteine from DL-2-Amino-Δ^2-thiazolinone-4-carboxylate with 2-Amino-Δ^2-thiazolinone-4-carboxylate Hydrolase, *S*-Carbamoyl-L-cysteine Hydrolase, and 2-Amino-Δ^2-thiazolinone-4-carboxylate Racemase ... 116
3.3.5 Production of D- and L-Lactate from DL-2-Chloropropionate with L-2- and DL-2-Haloacid Dehalogenases ... 116
4 Enzymatic Synthesis of Optically Active Amino Acids from Achiral Substances ... 117
4.1 Synthesis of L-Amino Acids with Ammonia Lyases ... 117
4.1.1 L-Aspartate Production from Fumarate and Ammonia with Aspartase ... 117
4.1.2 L-Phenylalanine Production from *trans*-Cinnamate and Ammonia with Phenylalanine Ammonia Lyase ... 118
4.2 L-Amino Acid Production from α-Keto Acids and L-Aspartate with Transaminases ... 118
4.3 L-Serine Production from Glycine with Serine Hydroxymethyltransferase ... 119
4.4 L-Amino Acid Production from α-Keto Acids and Ammonia with Amino Acid Dehydrogenases ... 119

4.4.1 L-Leucine Production from the α-Keto Analogue and Ammonia with Leucine Dehydrogenase ... 119
4.4.2 L-Alanine Production from Pyruvate and Ammonia with Alanine Dehydrogenase . 120
5 Enzymatic Synthesis of Stereospecifically Isotope-labeled Amino Acids, Amines, and Coenzymes ... 122
5.1 Synthesis of Stereospecifically Isotope-labeled Amines and Amino Acids with Amino Acid Decarboxylases ... 122
5.2 Synthesis of Stereospecifically Isotope-labeled Amino Acids and Pyruvate with Enzymes Catalyzing β-Replacements and α,β-Eliminations ... 122
5.3 Synthesis of Stereospecifically Isotope-labeled Pyridoxamine with Transaminases ... 125
5.4 Synthesis of Stereospecifically Isotope-labeled NAD(P)H with Amino Acid Dehydrogenases ... 125
5.5 Deuterium or Tritium-labeling of Amino Acids with Methionine γ-Lyase ... 126
6 Conclusion ... 127
7 Abbreviations ... 127
8 References ... 127

Various enantiomeric amino acids and α-hydroxy acids are produced by means of highly stereospecific microbial enzymes from optically active, racemic, and achiral starting materials which are produced cheaply by chemical industry. Several enzymes are used for the stereoselective labeling of amino acids, amines, and coenzymes with deuterium or tritium. The mechanisms of these enzyme reactions also are described here.

1 Introduction

Enzymes are characterized by causing reaction rate acceleration and by their specificity. Their enormous rate accelerations are clearly shown by the following example: the rate constant of nonenzymatic hydrolysis of urea is $4.15 \times 10^{-5}\ s^{-1}$ at 100 °C while that of urease-catalyzed hydrolysis is $3 \times 10^{4}\ s^{-1}$ at 20 °C. Another and more salient feature of enzymatic catalysis is the high specificity for substrates and reactions. Specificity can be absolute or nearly absolute in that some enzymes act only on substantially a single substrate molecule while others can act also on broad structural types such as on whole classes of molecules sharing a common structural denominator. As a rule, enzymes show a high reaction specificity, but some catalyze reactions apparently different from the inherent reactions, e.g., most hydrolases catalyze transferase reactions. In addition, another group of enzymes catalyze various types of reactions, i.e., they show a multicatalytic function.

Enzymes generally demonstrate unerring and complete stereospecificity in catalysis when they act on bonds linked to a chiral center of substrates or form chiral products from achiral substrates. Racemases and isomerases are exceptional in this respect. Stereospecificity is absolute even for enzymes which show broad specificity in a substrate structure; for example, D-amino acid oxidase (E.C. 1.4.3.3) catalyzes the oxidation of a variety of D-amino acids and D-imino acids, but does not oxidize L-amino acids at all; L-amino acid oxidase (E.C. 1.4.3.2) catalyzes the oxidation of L-amino acids exclusively [1, 2]. Stereospecificity is not strict for enzymes acting on groups distant from a chiral center of a substrate. Carboxylesterase (E.C. 3.1.1.1) catalyzes the hydrolysis of L- and D-tyrosine ethyl esters at the same rate [3], and *Pseudomonas* glutaminase A (E.C. 3.5.1.2) hydrolyzes L- and D-glutamine with a ratio of 1 to 0.65 [4]. Enzymes in some cases show unnecessarily high stereospecificity from the viewpoint of inherent catalysis as shown in the following examples: Most amino acid decarboxylases catalyze the decarboxylation of L-amino acids with retention of the substrate configuration; a proton is added to the carbon from the same direction from which the carboxyl group is removed [5]. In NAD(P)-dependent dehydrogenase reactions, only one of the two prochiral C-4 hydrogens of the nicotinamide ring of the coenzyme is transferred to the substrates [6]. As a whole, such high stereospecificity is difficult to attain by chemical catalysis, though asymmetric synthesis has been markedly developed in recent years [7].

In this review, we describe enzymatic synthesis of optically active compounds, in particular L- and D-amino acids and α-hydroxy acids, though, recently, a review on microbial and enzymatic production of amino acids has been published [8]. Enzymatic syntheses of optically active compounds can be classified from the standpoint of starting materials into the following categories (Fig. 1). (I) Optically active starting materials are converted into more useful optically active products with retention of optical activity, or with inversion of configuration. (II) Racemic compounds are converted to enantiomerically pure products. Enzymatic optical resolution is categorized here. (III) Achiral compounds are converted into optically active compounds by a single enzyme reaction or by a coupling of the reactions. In addition, the stereochemistry of hydrogen transfer, and stereospecific labeling of amino acids, amines, and coenzymes with isotopes also are described.

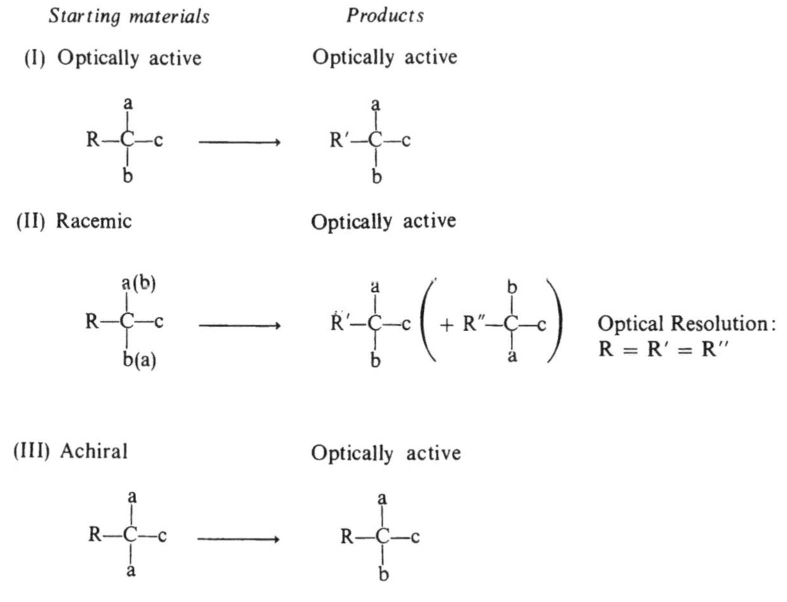

Fig. 1. Enzymatic synthesis of optically active compounds

2 Enzymatic Synthesis of L-Amino Acids and α-Hydroxy Acids from Optically Active Starting Materials

In this Section, the conversion of optically active substances into more useful optically active compounds with a single enzyme is described (Category I). Most of the enzymes used for this purpose are pyridoxal 5′-phosphate (PLP)-dependent enzymes. PLP enzymes play a central role in amino acid metabolism by catalyzing transamination, decarboxylation, racemization, replacement, addition, elimination, and a few other reactions of amino acids. There has been a recent review on their functions, structures, and reaction mechanisms [9].

2.1 Synthesis of *S*-Substituted L-Homocysteines with L-Methionine γ-Lyase

L-Methionine γ-lyase (EC 4.4.1.11), found in strains of *Chlostridium* and *Pseudomonas* [10], inherently catalyzes the conversion of L-methionine into methanethiol, α-ketobutyrate, and ammonia. The enzyme was purified to homogeneity and crystallized from *Pseudomonas putida* (= *ovalis*). It has a molecular weight of about 173,000 [10, 11] and consists of four identical subunits, each of which contains one mole of PLP as a Schiff base between the 4′-formyl group of the coenzyme and an ε-amino group of a lysine residue of the enzyme protein. In addition to α,γ-elimination (Eq. 1), L-methionine γ-lyase catalyzes γ-replacement (Eq. 2) of homocysteine derivatives, α,β-elimination (Eq. 3) and β-replacement (Eq. 4) of cysteine derivatives, and deamination (Eq. 5) and γ-addition (Eq. 6) of vinylglycine.

$$RSCH_2CH_2CH(NH_2)COOH \xrightarrow{H_2O} RSH + CH_3CH_2COCOOH + NH_3 \quad (1)$$

$$RSCH_2CH_2CH(NH_2)COOH \xrightarrow{R'SH} RSH + R'SCH_2CH_2CH(NH_2)COOH \quad (2)$$

$$RSCH_2CH(NH_2)COOH \xrightarrow{H_2O} RSH + CH_3COCOOH \quad (3)$$

$$RSCH_2CH(NH_2)COOH \xrightarrow{R'SH} RSH + R'SCH_2CH(NH_2)COOH \quad (4)$$

$$CH_2{=}CHCH(NH_2)COOH \xrightarrow{H_2O} CH_3CH_2COCOOH + NH_3 \quad (5)$$

$$CH_2{=}CHCH(NH_2)COOH \xrightarrow{RSH} RSCH_2CH_2CH(NH_2)COOH \quad (6)$$

vinylglycine

In Table 1, the ability of L-methionine γ-lyase to catalyze the elimination reaction with various amino acids is presented. In addition to L-methionine, which is the preferred substrate, several derivatives of L-methionine and L-cysteine are effective substrates, whereas D-methionine and D-cysteine are inert.

L-Methionine γ-lyase also catalyzes γ- and β-replacement reactions. L-Ethionine and S-ethyl-L-cysteine are synthesized from L-methionine and L-cysteine, respectively, and ethanethiol. Successive addition of enzyme and ethanethiol is very effective

Table 1. Substrate specificity (elimination) of L-methionine γ-lyase from *Pseudomonas putida*

Amino acids	Rel. act.
L-Methionine	100
L-Ethionine	90
DL-Methionine sulfone	87
DL-Methionine sulfoxide	35
L-Methionine-DL-sulfoximine	27
S-Methyl-L-methionine	8
L-Homocysteine	71
L-Cysteine	11
S-(β-Aminoethyl)-L-cysteine	14
S-Methyl-L-cysteine	67

Inert: D-methionine, L-norleucine, DL-α-aminobutyrate, L-cystathionine, L-norvaline, D-cysteine, and L-cystine

giving a yield of more than 95%. L-Ethionine is also produced by a γ-replacement reaction between the derivatives of methionine (e.g., homocysteine, methionine sulfone, and methionine sulfoxide) and ethanethiol. When ethanethiol is replaced by an alkanethiol (C_3–C_7) or an arylthioalcohol in the reaction system, a new sulfur-containing amino acid, corresponding to the thiol employed, is enzymatically synthesized (Table 2). Alkanethiols with longer carbon chains (C_8–C_{10}), however, are inert as an S-substituent donor. Substituted methanethiol and β-substituted ethanethiol also react with L-methionine giving the corresponding S-substituted homocysteines (Table 3).

Vinylglycine (2-amino-3-butenoate) is regarded as a common key intermediate in α,γ-elimination and γ-replacement reactions [12]. L-Methionine γ-lyase catalyzes the deamination of vinylglycine to produce α-ketobutyrate and ammonia [13]. It also catalyzes the γ-addition of various alkanethiols to vinylglycine to yield the corresponding S-substituted homocysteines [13]. The relative activity of the enzyme for various alkanethiols in the γ-addition reaction of vinylglycine is close to that obtained for the γ-replacement reactions of methionine (Table 4). The mechanism for the enzyme reactions in α,γ-elimination, γ-replacement, and γ-addition is shown in Fig. 2 [13]. A sulfur amino acid substrate reacts with the enzyme-bound PLP to form an aldimine Schiff base (I), and an α-hydrogen of the substrate is lost due to the electron attractivity of the protonated pyridine-N and imine-N. Thus, a quinoid intermediate of S-substituted homocysteine (III) is formed, and a β-hydrogen and a γ-substituent are released in the intermediate (γ-elimination). The resulting β,γ-unsaturated compound (IV) is regarded as a quinoid intermediate of vinylglycine, which is derived from the PLP Schiff base of the compound (II). Intermediate (IV) undergoes γ-addition (or γ-replacement) of a thiol to form product (VII) through an intermediate (V) (course *a*), or alternatively tautomerization to form the Schiff base of aminocrotonate (VI) (course *b*). The intermediate is hydrolyzed to yield aminocrotonate, which is spontaneously converted into α-ketobutyrate and ammonia through α-iminobutyrate (α-elimination). Based on this mechanism various *S*-substituted derivatives of L-homocysteine are synthesized with the enzyme.

Table 2. Relative reactivity of thiols in the γ-replacement reactions with L-methionine γ-lyase

Thiols	Products		Thiols	Products	
Ethylmercaptan	$CH_3CH_2—S—R$	(100)[a]	*sec*-Pentylmercaptan	$CH_3(CH_2)_2CH(CH_3)—S—R$	(12)[a]
n-Propylmercaptan	$CH_3CH_2CH_2—S—R$	(81)	*tert*-Pentylmercaptan	$(CH_3)_3CCH_2—S—R$	(9)
iso-Propylmercaptan	$(CH_3)_2CH—S—R$	(38)	*n*-Hexylmercaptan	$CH_3(CH_2)_4CH_2—S—R$	(17)
n-Butylmercaptan	$CH_3(CH_2)_2CH_2—S—R$	(43)	*tert*-Hexylmercaptan	$(CH_3)_2C(S—R)(CH_2)_2CH_3$	(5)
iso-Butylmercaptan	$(CH_3)_2CHCH_2—S—R$	(75)	*n*-Heptylmercaptan	$CH_3(CH_2)_5CH_2—S—R$	(4)
sec-Butylmercaptan	$CH_3CH(CH_2CH_3)—S—R$	(27)	Cyclohexylmercaptan	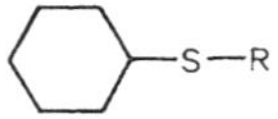	(10)
tert-Butylmercaptan	$(CH_3)_3C—S—R$	(26)	Benzylmercaptan	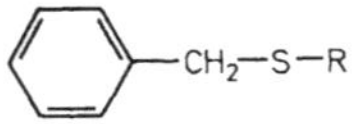	(71)
n-Pentylmercaptan	$CH_3(CH_2)_3CH_2—S—R$	(43)	β-Naphthylmercaptan	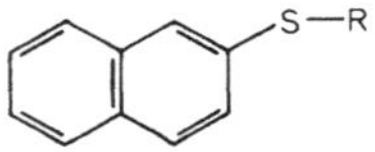	(6)
iso-Pentylmercaptan	$(CH_3)_2CHCH_2CH_2—S—R$	(52)			

[a] Relative activity. R: $—CH_2CH_2CH(NH_2)COOH$
Inert: *tert*-heptyl, *n*-octyl, *tert*-octyl, *tert*-nonyl, and *n*-decylmercaptans

One of the characteristics of the enzyme is its reactivity for selenium derivatives of methionine [14]. It catalyzes the α,γ-elimination of selenomethionine to yield α-ketobutyrate, ammonia, and methaneselenol, and also the γ-replacement reactions of selenomethionine with various thiols to produce *S*-substituted homocysteines and methaneselenol. In addition, L-methionine and its derivatives react with various selenols to form the corresponding *Se*-substituted selenohomocysteines, although selenols are less efficient γ-substituent donors than thiols. Vinylglycine serves as a substrate in the γ-addition reaction also with selenols to form the corresponding *Se*-substituted selenohomocysteines. Selenium amino acids, in particular optically active ones, are synthesized chemically with little success, because of unstability of the element. The enzymatic replacement and addition reactions facilitate the synthesis of various selenium amino acids, even if labeled with ^{14}C, ^{75}Se, and ^{3}H.

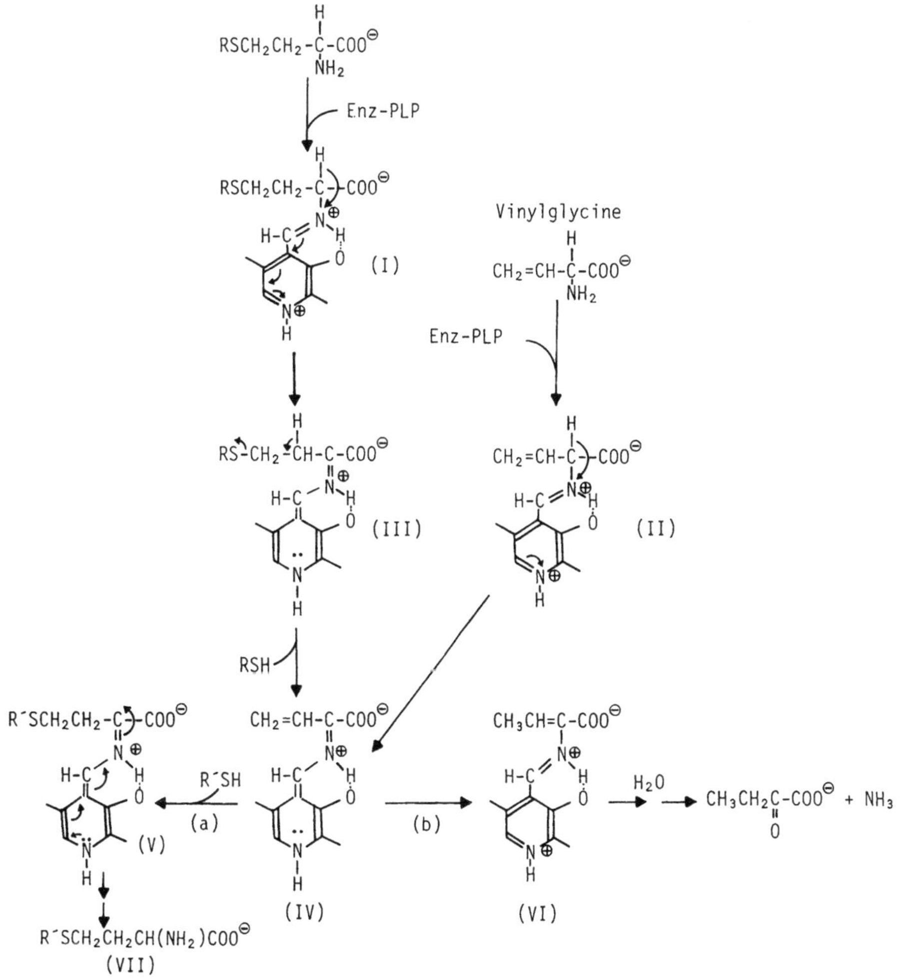

Fig. 2. Reaction mechanism of methionine γ-lyase. Enz= apoenzyme

Table 3. Reactivity of substituted methane and ethanethiols in γ-replacement reactions with L-methionine γ-lyase

Thiols	Products	Rel. act.
CH_3CH_2SH	$CH_3CH_2{-}S{-}R$	100
$HOOCCH_2SH$	$HOOCCH_2{-}S{-}R$	11
$CH_3CH_2OOCCH_2SH$	$CH_3CH_2OOCCH_2{-}S{-}R$	84
$HOCH_2CH_2SH$	$HOCH_2CH_2{-}S{-}R$	111
$HOOCCH_2CH_2SH$	$HOOCCH_2CH_2{-}S{-}R$	7
$H_2NCH_2CH_2SH$	$H_2NCH_2CH_2{-}S{-}R$	44
$H_3COCNHCH_2CH_2SH$	$H_3COCNHCH_2CH_2{-}S{-}R$	267

R: $-CH_2CH_2CH(NH_2)COOH$

Table 4. Reactivity of thiols in the γ-addition reaction of vinylglycine and γ-replacement reaction of L-methionine catalyzed by L-methionine γ-lyase

Thiols	Vinylglycine	Methionine	Products
Ethanethiol	100[a]	100	$CH_3CH_2{-}S{-}R$
n-Propanethiol	82	81	$CH_3CH_2CH_2{-}S{-}R$
n-Hexanethiol	23	17	$CH_3(CH_2)_4CH_2{-}S{-}R$
α-Toluenethiol	72	71	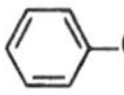 $-CH_2{-}S{-}R$

[a] Relative rate of the reaction with L-methionine is 75%.
R: $-CH_2CH_2CH(NH_2)COOH$

2.2 Synthesis of L-Tyrosine and its Related Compounds with β-Tyrosinase

β-Tyrosinase[L-tyrosine phenol-lyase (deaminating)] (E.C. 4.1.99.2) catalyzes the α,β-elimination of L-tyrosine to produce pyruvate, phenol, and ammonia. The enzyme has been crystallized from *Citrobacter intermedius* (*Escherichia intermedia*) and from *Erwinia herbicola* [15,16]. The *Erwinia* enzyme has a molecular weight of approximately 259,000 and contains two moles of PLP per mole of enzyme. Yamada and his coworkers have established optimal cultivation conditions of the bacteria for maximum production of β-tyrosinase [17]. The enzyme is produced inducibly by addition of L-tyrosine to the medium. Although L-phenylalanine is not an inducer, it cooperatively enhances the effect of L-tyrosine. This is probably due to the inhibition of enzyme action by L-phenylalanine. A decrease in the rate of phenol formation causes an increase in enzyme production; phenol inhibits both bacterial growth and the formation of β-tyrosinase. The concentration of the enzyme can reach up to 10% of the total soluble protein in *Erwinia herbicola* under optimal conditions.

In addition to the α,β-elimination reaction, the enzyme catalyzes a β-replacement reaction and a reverse reaction of α,β-elimination [15,18]. In the α,β-elimination reaction, D-tyrosine, L- and D-serine, S-methyl-L-cysteine, and β-chloro-L-alanine act as substrates besides L-tyrosine. L-Tyrosine is formed by the β-replacement reaction between the substrates for α,β-elimination and phenol. The D-isomers of amino

Table 5. Synthesis of L-tyrosine-related amino acids from pyruvate, NH_3, and phenols by β-tyrosinase

Phenols	L-Amino acids produced	
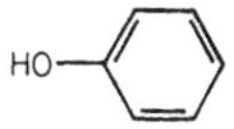	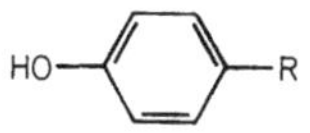	(100)[a]
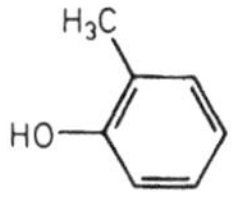	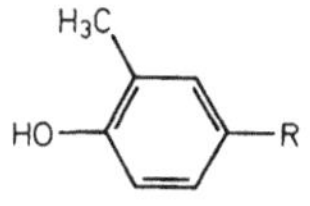	(45.4)
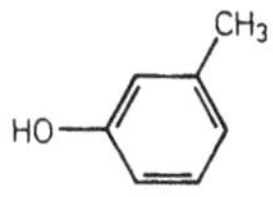	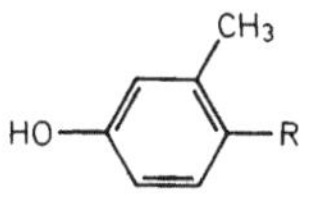	(52.7)
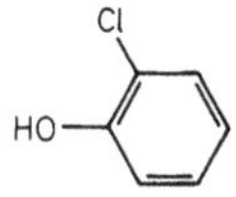	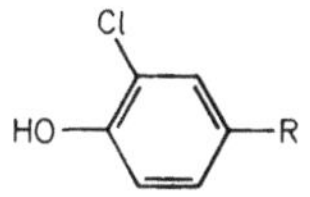	(47.3)
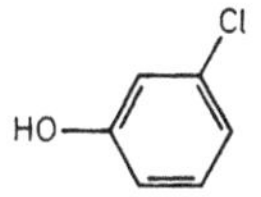	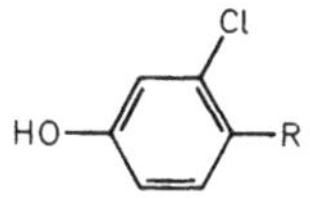	(32.7)
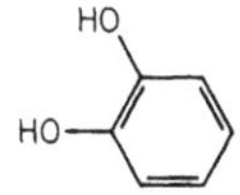	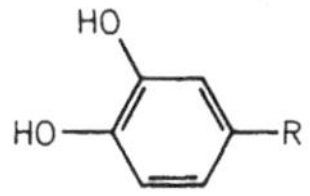	(60.0)
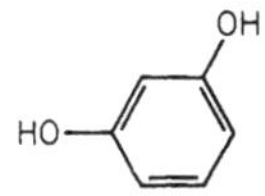	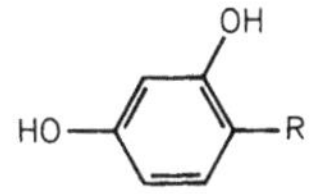	(58.2)

[a] Relative velocity.
R: $-CH_2CH(NH_2)COOH$

acids substrates also give tyrosine derivatives. When phenol is replaced by pyrocatechol, resorcinol, pyrogallol, and hydroxyhydroquinone, 3,4-dihydroxy-L-phenylalanine (L-DOPA), 2,4-dihydroxy-L-phenylalanine, 2,3,4-trihydroxy-L-phenylalanine, and 2,4,5-trihydroxy-L-phenylalanine are synthesized, respectively. For example, 5.5 g of L-DOPA is obtained per 100 ml of reaction mixture [19]. The enzyme catalyzes also the reverse reaction of α,β-elimination, i.e., the formation of L-tyrosine from phenol, pyruvate, and ammonia [20]. More than 6.0 g of L-tyrosine per 100 ml of reaction mixture is synthesized. When various derivatives of phenol are used as β-substituent donors, the corresponding tyrosine derivatives are produced (Table 5). Ikeda et al. immobilized the enzyme on Sepharose in order to synthesize L-tyrosine continuously [21]. *Erwinia herbicola* cells immobilized in collagen matrices are more resistant to heat, contact with phenolic compounds, and to changes in pH than the intact cells [22].

The mechanism of the multifunctional β-tyrosinase reaction is shown in Fig. 3. The reaction mechanism is analogous to that for α,γ-elimination and γ-addition

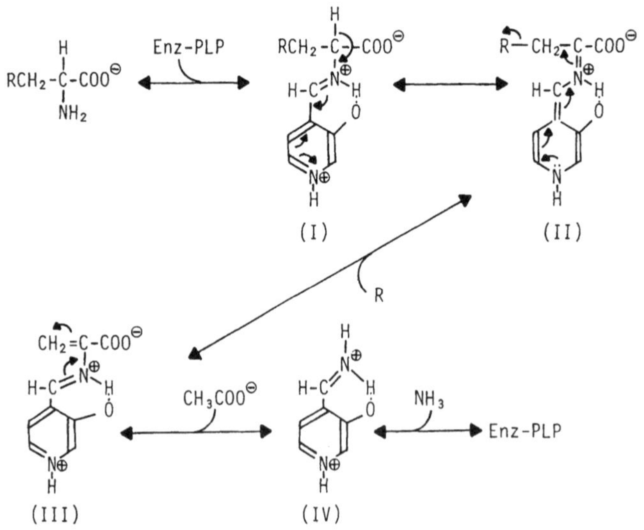

Fig. 3. Reaction mechanism of β-tyrosinase. Enz = apoenzyme

reactions by L-methionine γ-lyase described above. Thus, the quinoid intermediate (II) is formed through an aldimine Schiff base of the substrate and PLP (I), and a β-substituent is released in the intermediate. From the resulting Schiff base of amino-acrylate (III), pyruvate and then ammonia are released. The PLP-α-aminoacrylate complex (III) is a common key intermediate for all the α,β-eliminations, β-replacements, and the reverse reactions [12]. Kinetic studies show that the reverse reaction proceeds via an ordered *Ter-Uni* mechanism, and that pyruvate may be the second substrate to combine with the enzyme [23]. Spectrophotometric studies indicate that ammonia is the first substrate that interacts with bound PLP [24]. This mechanism has been further confirmed by the substrate proton-exchange reaction and by spectral analysis [25]. The β-replacement reaction between L-tyrosine and resorcinol proceeds with retention of configuration at the β-carbon [26]. The replacement of the β-substituent by a proton in the α,β-elimination of L-serine and L-tyrosine also occurs with retention of configuration [27].

2.3 Synthesis of L-Cysteine-related Amino Acids and Their Selenium Analogues with Tryptophan Synthase

Tryptophan synthase [L-serine hydro-lyase (adding indoleglycerolphosphate)] (E.C. 4.2.1.20) is found in various bacteria, yeast, molds, and plants, and catalyzes reactions 1–3 shown in Table 6 [28]. Reaction 3 is physiological, and reactions 1 and 2 are regarded as partial reactions of reaction 3. Tryptophan synthase of *Escherichia coli* is composed of two kinds of subunits, two α-subunits and one β_2-dimer [28]. One mole of PLP binds per mole of the β-subunit. The crystalline $\alpha_2\beta_2$-complex was obtained after a six-fold purification from the strain trp $R^-\Delta$trp ED102/F'Δtrp ED/102 of *E. coli* developed by Yanofsky et al. [28]. In this strain about 16% of the total soluble protein is the tryptophan synthase complex.

Table 6. Typical reactions catalyzed by *Escherichia coli* tryptophan synthase and its subunits

Reactions	Subunits involved
1. Indole-3-glycerol phosphate ⇄ Indole + D-Glyceraldehyde-3-phosphate	α, $\alpha_2\beta_2$
2. Indole + L-Serine ⇄ L-Tryptophan + H_2O	β_2, $\alpha_2\beta_2$
3. Indole-3-glycerol phosphate + L-Serine ⇄ L-Tryptophan + D-Glyceraldehyde-3-phosphate + H_2O	$\alpha_2\beta_2$

Each of the subunits catalyzes the reactions shown in Table 6. The α-subunit catalyzes an aldol cleavage reaction of indole-3-glycerol phosphate to form indole and D-glyceraldehyde-3-phosphate, and then L-tryptophan is formed by a β-replacement reaction between indole and L-serine catalyzed by the β_2-dimer. During these processes indole formed by the α-subunit is not released from the enzyme. The $\alpha_2\beta_2$-complex does not catalyze the elimination reaction of tryptophan at all. L-Tryptophan, therefore, is produced with a high yield by reactions 1 and 2.

Tryptophan synthase also catalyzes the β-replacement reaction of L-serine with various thiols to produce S-substituted L-cysteines (Table 7) [29]. L-Serine is successfully replaced by various other L-amino acids in the β-replacement reaction (Table 8) [29]. L-Threonine and L-vinylglycine react with toluenethiol to form the corresponding β-methyl-S-benzylcysteine. In the β-replacement reactions various selenols such as tolueneselenol and methaneselenol also react with L-serine to form the corresponding derivatives of Se-L-cysteine [30]. The reaction mechanism of tryptophan synthase is essentially the same as that of β-tyrosinase and methionine γ-lyase described above [28].

Table 8. Reactivity of β-substituent acceptors in the β-replacement reaction with benzylmercaptan catalyzed by tryptophan synthase

Amino acids	Rel. act.
L-Serine	100
O-Methyl-DL-serine	41
β-Cyclo-L-alanine	35
O-Acetyl-L-serine	13
S-Methyl-L-cysteine	24
S-Ethyl-L-cysteine	12
Se-Methyl-DL-seleocysteine	84
L-Cysteine	0
L-Threonine	1
L-*allo*-Threonine	0
L-Vinylglycine	0.5

Table 7. Synthesis of S-substituted L-cysteines from L-serine and thiols by tryptophan synthase

Thiols	Products		Thiols	Products	
Benzylmercaptan	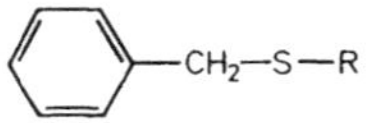	(100)[a]	*iso*-Butylmercaptan	$(CH_3)_2CHCH_2{-}S{-}R$	(40)[a]
Methylmercaptan	$CH_3{-}S{-}R$	(19)	*n*-Pentylmercaptan	$CH_3(CH_2)_3CH_2{-}S{-}R$	(37)
Ethylmercaptan	$CH_3CH_2{-}S{-}R$	(21)	*n*-Hexylmercaptan	$CH_3(CH_2)_4CH_2{-}S{-}R$	(19)
n-Propylmercaptan	$CH_3CH_2CH_2{-}S{-}R$	(152)	*n*-Heptylmercaptan	$CH_3(CH_2)_5CH_2{-}S{-}R$	(12)
n-Butylmercaptan	$CH_3(CH_2)_2CH_2{-}S{-}R$	(113)	*n*-Octylmercaptan	$CH_3(CH_2)_6CH_2{-}S{-}R$	(3)
sec-Butylmercaptan	$CH_3CH(CH_2CH_3){-}S{-}R$	(24)	Phenylmercaptan	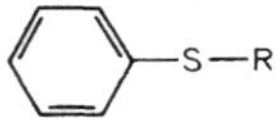	(8)

[a] Relative activity. R: $-CH_2CH(NH_2)COOH$
Inert: *tert*-butyl and *n*-decylmercaptan

2.4 Synthesis of Amino Acids and Their Derivatives with Other PLP-enzymes

L-Tryptophan and its derivatives are synthesized with tryptophanase [L-tryptophan indole-lyase (deaminating)] (E.C. 4.1.99.1). The enzyme is found widely in bacteria, and has been purified to characterize enzymologic and physicochemical properties [31, 32]. It catalyzes α,β-elimination, β-replacement, and the reverse reaction of α,β-elimination in a similar way to β-tyrosinase. In addition to L-tryptophan, L-cysteine, S-methyl-L-cysteine, β-chloro-L-alanine, and L-serine serve as substrates for the α,β-elimination and the β-replacement reactions [33]. L-Tryptophan is synthesized from indole, pyruvate, and ammonia by the reverse reaction of α,β-elimination [32]. Various indole derivatives serve as substrates for the synthesis of the corresponding L-tryptophan derivatives (Table 9).

L-Homocysteine and L-selenohomocysteine are synthesized with *O*-acetylhomoserine sulfhydrylase [*O*-acetylhomoserine (thiol)-lyase] (EC. 4.2.99.10) [34]. The enzyme catalyzes γ- and β-replacement reactions, but does not lead to eliminations at all. Thus it is similar to tryptophan synthase. The purified enzyme from *Saccharomyces cerevisiae* catalyzes the γ-replacement reaction between *O*-acetylhomoserine

Table 9. Synthesis of L-tryptophan analogues from pyruvate, NH_3, and corresponding indoles by tryptophanase

Indole and its derivatives	L-Amino acids synthesized	
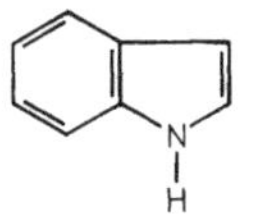	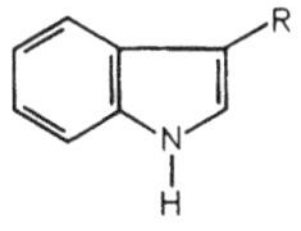	(100)[a]
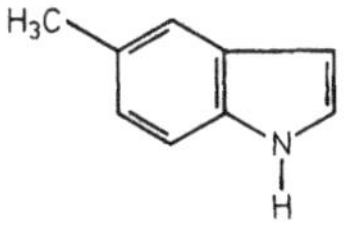	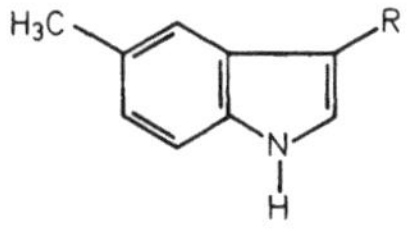	(60.7)
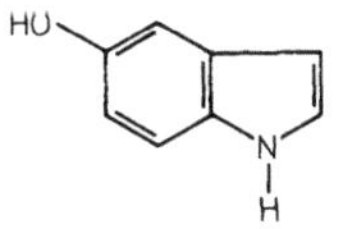	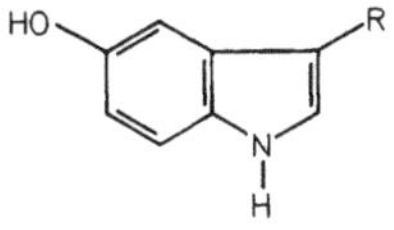	(39.6)
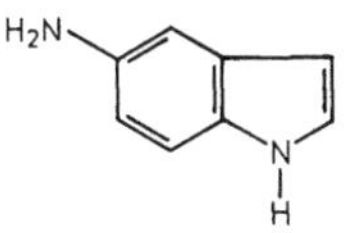	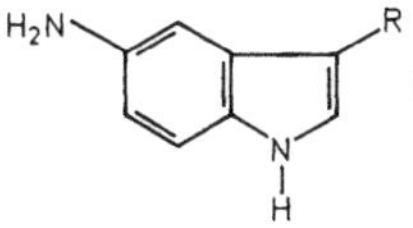	(19.6)

[a] Relative velocity.
R: $-CH_2CH(NH_2)COOH$
Inert: 5-Methylindole, 5-carboxyindole, 2-phenylindole, and *N*-methylindole

and hydrogen sulfide, yielding L-homocysteine [34]. Hydrogen selenide also serves as a substituent donor in the reaction, and consequently L-selenohomocysteine is produced. *Se*-L-Cysteine is also synthesized by a β-replacement reaction between hydrogen selenide and L-serine *O*-sulfate [35].

L-Cysteine- and Se-L-cysteine-related amino acids are synthesized with cystathionine β-synthase (EC 4.2.1.22) [35,36]. The enzyme is found in mammals and microorganisms. The purified enzyme from *Saccharomyces cerevisiae* catalyzes the β-replacement reaction between L-serine and thiols or selenols, giving the corresponding sulfur- or selenium-amino acids [36].

D-Cysteine is synthesized by the β-replacement reaction between β-chloro-D-alanine and hydrogen sulfide catalyzed by β-chloro-D-alanine hydrogen chloride lyase [37]. The enzyme was found in *Pseudomonas putida* CR 1-1 and purified from the cell extract. It catalyzes α,β-elimination of β-chloro-D-alanine to form pyruvate, ammonia, and hydrogen chloride, and a β-replacement reaction [38]. Nagasawa et al. have studied the relationship between the cultivation conditions of *P. putida* CR 1-1 and D-cysteine production with resting cells [39]. Under optimum conditions, 20.6 mg of D-cysteine per ml of the reaction mixture was produced from β-chloro-D-alanine with a molar yield of 100%.

L-Alanine is produced from L-aspartate with aspartate β-decarboxylase (L-aspartate 4-carboxy-lyase) (EC 4.1.1.1). The enzyme catalyzes the removal of the 4-carboxyl group of L-aspartate to form L-alanine. It has been purified and crystallized from *Alcaligenes faecalis* und *Pseudomonas dacunhae* [40,41]. L-Aspartate is produced industrially from fumarate and ammonia with bacterial aspartase as described below. Chibata et al. have established an industrial method for the production of L-alanine (400 g l^{-1}) from L-aspartate with intact cells of *P. dacunhae* [42].

2.5 Synthesis of D-2-Hydroxy Acids with L-2-Haloacid Dehalogenase

L-2-Haloacid dehalogenase (EC 3.8.1.2) catalyzes dehalogenation of L-2-haloalkanoic acids to yield D-2-hydroxy acids, e.g., D-2-lactate is produced from L-2-chloropro-

$$\text{L-RCH(X)COOH} + H_2O \rightarrow \text{D-RCH(OH)COOH} + HX$$

X: halogen atom

pionate. Thus, the enzymatic reaction proceeds with inversion of configuration. The enzyme was purified from *Pseudomonas putida* cells grown in a medium containing DL-2-chloropropionate [43]. It has a molecular weight of 34,000. As shown in Table 10, various 2-haloalkanoic acids serve as substrates of the enzyme, producing the corresponding 2-hydroxy acids. Although achiral haloacetic acids are good substrates of the enzyme, the reaction is strictly stereospecific; D-isomers are inert. 2,2-Dihaloalkanoic acids also are dehalogenated to produce the corresponding 2-oxoalkanoic acids.

Table 10. Substrate specificity of L-2-haloaciddehalogenase

Haloacids	Rel. act.
Monobromoacetic acid	356
Monoiodoacetic acid	255
Monochloroacetic acid	184
Dichloroacetic acid	30
DL-2-Bromopropionic acid	247
L-2-Chloropropionic acid	100
2,2-Dichloropropionic acid	10
DL-2-Bromobutyric acid	98
DL-3-Chlorobutyric acid	1

Inert: Monofluoroacetic acid, trichloroacetic acid, chloroacetamide, chloroacetaldehyde, D-2-chloropropionic acid, 3-chloropropionic acid, DL-2-bromobutyric acid, and DL-bromo-*iso*-butyric acid

3 Enzymatic Synthesis of Optically Active Amino Acids and α-Hydroxy Acids from Racemic Starting Materials

The chiral compounds, in general, are produced chemically as a racemate. Therefore, optical resolution is required to obtain the enantiomers. Several methods have been developed for the synthesis of one enantiomer from the corresponding racemates, and may be classified into four categories: physicochemical methods, chemical methods, enzymatic methods, and microbial methods [44]. The physicochemical methods include mechanical separation, selective crystallization, replacing crystallization, and chromatographic separation [45,46]. In the chemical resolution, the racemates or their derivatives are treated with chiral compounds, and one of the diastereoisomers is fractionally separated from the other by means of crystallization, chromatographic procedures, and others [47]. Although both methods are useful for stereoselective analysis, they are not suitable for industrial resolution. At present, the most useful and convenient methods for optical resolution are those employing enzymes (or microbial cells) which rely on the structural and stereochemical substrate specificity of enzymes. There are three different methods: (1) asymmetric derivatization of racemates with enzymes, (2) stereospecific hydrolysis of racemates, and (3) stereospecific derivatization of racemates by a combination of multiple enzymes which show a different stereospecificity.

3.1 Synthesis of L- and D-Amino Acids by Asymmetric Derivatization of Racemates

In this method one enantiomer of a racemate is stereospecifically derivatized by enzymes [48,49]. For example, chemically synthesized *N*-acetyl-DL-amino acids are incubated with aniline in the presence of papain (EC 3.4.22.2) to form insoluble *N*-acetyl-L-amino anilides, whereas the *N*-acetyl-D-amino acids remain in solution because of their insusceptibility to the enzyme. *N*-Acetyl-L-amino acid anilides thus purified are hydrolyzed chemically to give the L-amino acids. The N acetyl-D-amino acids are separately hydrolyzed to D-amino acids.

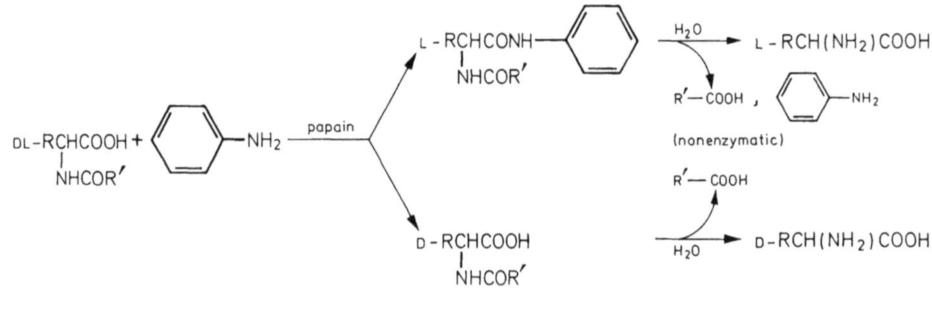

3.2 Synthesis of L- and D-Amino Acids by Stereospecific Hydrolysis of Racemates

3.2.1 Production of Amino Acids from *N*-Acyl-DL-amino Acids with Aminoacylases

Aminoacylase (EC 3.5.1.14) catalyzes the hydrolysis of various *N*-acyl-L-amino acids, such as *N*-acetyl-, *N*-chloroacetyl- and *N*-propylderivatives of L-amino acids. The enzyme has been found in hog kidney, *Aspergillus*, and *Penicillium*, and requires Co^{2+} for maximum activity [50-52]. The substrate specificity of the enzyme, which

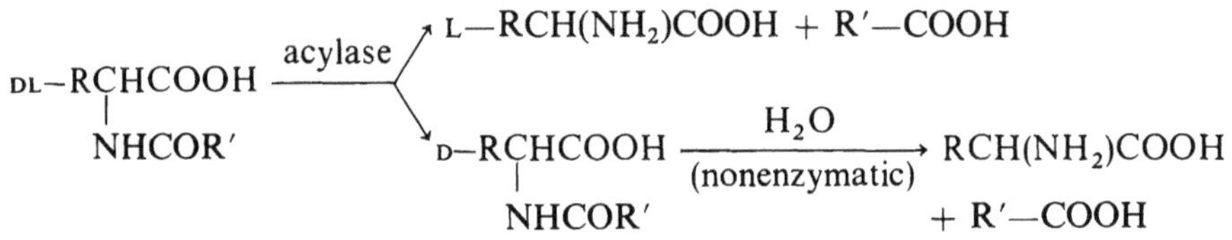

was partially purified from *Aspergillus sojae*, is shown in Table 11. The fungal enzyme catalyzes the hydrolysis of the *N*-acyl derivatives of aromatic L-amino acids effectively, and has been used industrially for the resolution of tryptophan and phenylalanine racemates. When *N*-acetyl-DL-tryptophan is incubated with the fungal aminoacylase, *N*-acetyl-L-tryptophan is selectively hydrolyzed to L-tryptophan, which is then crystallized from the solution. *N*-Acetyl-D-tryptophan in the mother liquor is racemized with acetic anhydride, and the racemate is again used as a starting material. Tosa et al. have immobilized the acylase on DEAE-Sephadex, and have developed a continuous resolution method, which is now utilized in industry [53]. There are also other amino acid acylase: ε-*N*-acyl-L-lysine amino acylase (EC 3.5.1.17) and *N*-acetyl-D-amino acid aminoacylase (EC class 3.5.1.) [54-56]. These enzymes also are applicable to industrial optical resolution of their substrate amino acids.

Table 11. Substrate specificity of aminoacylase from *Aspergillus oryzae*

Amino acids	Acyl group	Rel. act.
Methionine	Acetyl	100
	Chloroacetyl	180
	Glycyl	26
Alanine	Acetyl	73
	Chloroacetyl	144
Valine	Acetyl	25
Leucine	Acetyl	31
	Chloroacetyl	46
Isoleucine	Acetyl	65
Alloisoleucine	Acetyl	4.7
Threonine	Chloroacetyl	14
Glutamate	Acetyl	37
	Chloroacetyl	53
Aspartate	Acetyl	10
	Chloroacetyl	10
Lysine	α,ε-Diacetyl	0
	ε-Benzoyl-α-acetyl	83
	ε-Benzoyl-α-chloroacetyl	158
Tryptophan	Acetyl	74
	Chloroacetyl	63
Phenylalanine	Acetyl	145
	Chloroacetyl	215
Tyrosine	Chloroacetyl	226

3.2.2 Production of Amino Acids from DL-Hydantoins with Hydantoinase

Amino acid hydantoins are intermediary products in conventional amino acid synthesis by the Bucherer method [57]. Hydantoinase (dihydropyrimidinase) (EC 3.5.2.2) catalyzes the hydrolytic ring-opening reaction of 5-substituted hydantoins to produce *N*-carbamoyl-α-amino acids [58] (Fig. 4). This enzyme is widely found in microorganisms [59].

The crystalline enzyme from *Pseudomonas striata* has a molecular weight of about 190,000 and is composed of four identical subunits with a molecular weight of 48,000 [60]. The substrate specificity is listed in Table 12. Hydantoins derived from alanine, valine, leucine, methionine, phenylglycine, and hydroxyphenylglycine are

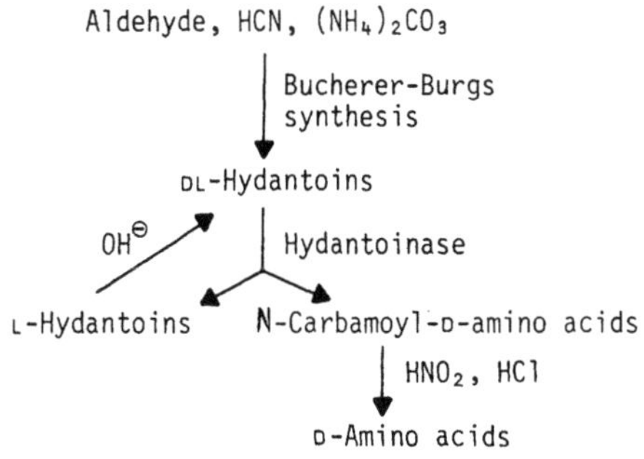

Fig. 4. Synthesis of D-amino acids from DL-hydantoins with hydantoinase

Table 12. Hydrolysis of hydantoins by *Pseudomonas striata* hydantoinase

5-Substituted hydantoins	Rel. act.
Dihydrouracil	100
H—R	13
CH_3-R	45
$(CH_3)_2CH-R$	15
$(CH_3)CHCH_2-R$	48
$CH_3SCH_2CH_2-R$	48
	25
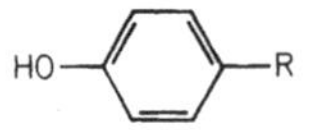	16
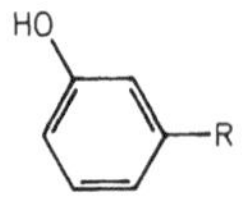	5
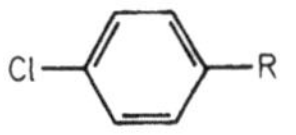	19
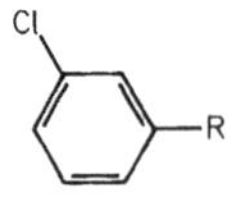	10
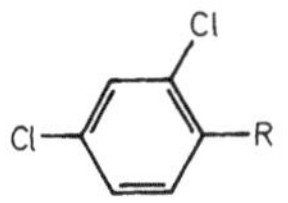	7
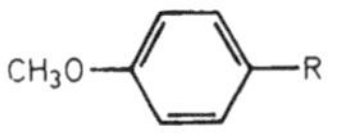	4

R: Hydantoin
Inert: diphenyl-, dimethyl-, carboxymethyl-, and aminobutylhydantoins

hydrolyzed. However, the hydantoins derived from aspartate, lysine, and 5,5-disubstituted hydantoins are inert. The *N*-carbamoyl amino acids produced have the same optical properties as the authentic *N*-carbamoyl D-amino acids derived from D-amino acids. *N*-Carbamoyl derivatives derived from 5-substituted hydantoins by hydantoinase can be transformed into D-amino acids by treatment with nitrous acids. Hydantoins are readily racemized in dilute alkaline solutions. Although the L-isomers of hydantoins are not hydrolyzed by the enzyme, all the DL-hydantoins used are converted to *N*-carbamoyl-D-amino acids which are decomposed to D-amino acids (Fig. 4).

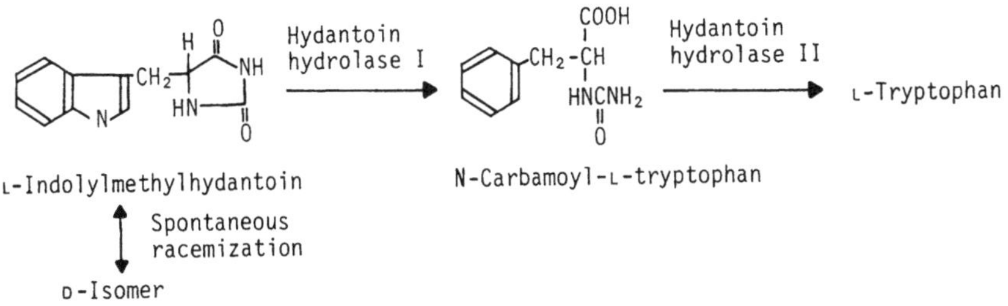

Fig. 5. Enzymatic synthesis of L-tryptophan from DL-5-indolylmethylhydantoin

An alkalophile, *Bacillus* sp., which forms hydantoinase abundantly, was isolated and the cells were immobilized in a polyacrylamide gel lattice [61]. N-Carbamoyl-D-hydroxyphenylglycine efficiently was produced with the cells from DL-5-*p*-hydroxyphenylhydantoin. The stability of hydantoinase from the cells of *Bacillus* sp. is markedly higher at alkaline pH than that of the *P. striata* cells. Thus, the alkalophilic bacterium is more suitable for the production of N-carbamoyl-D-*p*-hydroxylphenylglycine under alkaline conditions.

Sano et al. have isolated bacteria from soil which hydrolyze DL-5-indolylmethylhydantoin to L-tryptophan, and have developed a new enzymatic process to produce L-tryptophan with the bacterial cells [62]. The DL-5-indolylmethylhydantoin-hydrolyzing enzymes are inducibly formed in the bacterial cells by addition of DL-5-indolylmethylhydantoin to the medium. L-Tryptophan is produced in a molar yield of 82% with the intact cells and 100 mg ml^{-1} of DL-indolylmethylhydantoin. Coupling of two enzyme reactions results in the hydrolysis of DL-hydantoin to L-tryptophan (Fig. 5).

3.2.3 Production of Amino Acids from Amino Acid Esters with Esterases

Various amino acid esters are hydrolyzed stereospecifically by chymotrypsin and other proteolytic enzymes which show esterase activity [63].

$$\text{DL—RCH(NH}_2\text{)COOR}' \xrightarrow{\text{esterase}} \begin{cases} \text{L—RCH(NH}_2\text{)COOH} + \text{R}'\text{OH} \\ \text{D-RCH(NH}_2\text{)COOR}' \xrightarrow[\text{(nonenzymatic)}]{H_2O} \end{cases}$$

$$\text{L—RCH(NH}_2\text{)COOH} + \text{R}'\text{OH}$$

The resulting L-amino acids can be separated from the unreacted D-amino acid esters by extraction as they have different solubility in solvents. Amino acid esters may be hydrolyzed spontaneously to some extent. This leads to a decrease in the optical purity of L-amino acids produced by this method.

3.2.4 Production of Amino Acids from Amino Acid Amides with Amidases

L-Amino acid amides are hydrolyzed stereospecifically by amino acid amidases from mammalian kidney and pancreas, and from various microorganisms [64]. L-Amino acids thus formed are separated from the unreacted D-amino acid amide by the differences in solubility in various solvents. Since amino acid amides are less sus-

$$\text{DL-RCH(NH}_2\text{)CONH}_2 \xrightarrow{\text{amidase}} \begin{cases} \text{L-RCH(NH}_2\text{)COOH} + \text{NH}_3 \\ \text{D-RCH(NH}_2\text{)CONH}_2 \xrightarrow[\text{(nonenzymatic)}]{H_2O} \end{cases}$$

$$\text{D-RCH(NH}_2\text{)COOH} + \text{NH}_3$$

ceptible to spontaneous hydrolysis than the esters, amino acids with higher optical purity in general are produced by this method.

3.3 Synthesis of Amino Acids and α-Hydroxy Acids by Stereospecific Resolution of Racemates by a Combination of Multiple Enzymes

3.3.1 Production of D-α-Aminobutyrate and D-Methionine from DL-Methionine with Methionine γ-Lyase and D-Amino Acid Transaminase

DL-Methionine is chemically synthesized as a feed additive on a large scale in industry, and is one of the cheapest amino acids. Methionine γ-lyase catalyzes α,γ-elimination of L-methionine to produce α-ketobutyrate, methanethiol, and ammonia as described above. D-Methionine does not serve as a substrate and has no effect on the enzyme reaction. Therefore, D-methionine can be obtained by chromatographic separation from the reaction products after incubation of the enzyme with DL-methionine.

When the enzyme reaction is coupled with the reaction of D-amino acid transaminase (D-alanine transaminase) (EC 2.6.1.21), which was purified from *Bacillus sphaericus* [65], D-α-aminoburyrate and α-keto-γ-methylthiobutyrate are produced by the transamination between α-ketobutyrate and D-methionine. D-α-Aminobutyrate is a component

$$\text{L-Methionine} \xrightarrow[\searrow\ \text{CH}_3\text{SH, NH}_3]{\text{Methionine}\ \gamma\text{-lyase}} \alpha\text{-Ketobutyrate} \xrightarrow[\text{D-Methionine}\ \nearrow]{\text{D-Amino acid transaminase}} \text{D-}\alpha\text{-Aminobutyrate} + \alpha\text{-Keto-}\gamma\text{-methylthiobutyrate}$$

of an antibiotic, doricin (Fig. 6). At the equilibrium of the coupled reaction, DL-methionine is converted to an equimolar amount of D-methionine, D-α-aminobutyrate, α-ketobutyrate, and α-keto-γ-methylthiobutyrate, which are separated chromatographically from each other [66].

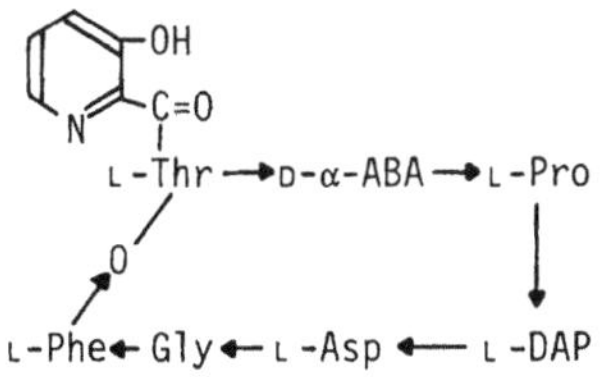

Fig. 6. Structure of doricin. D-α-ABA: D-α-aminobutyrate. L-DAP: L-*p*-dimethylamino-*N*-methylphenylalanine

3.3.2 Production of L-Tryptophan from Indole and DL-Serine with Tryptophan Synthase and Amino Acid Racemase

L-Tryptophan is synthesized from indole and L-serine by the catalysis of tryptophan synthase, which is abundantly produced by a mutant of *E. coli* as described above. L-Serine, however, is not cheaply available, but the racemate is. Amino acid racemase (EC 5.1.2.10) from *Pseudomonas putida* catalyzes racemization of serine, but not that of tryptophan [67]. When the tryptophan synthase reaction is coupled with the racemase reaction (both enzymes have similar pH optima), D-serine is converted to the L-enantiomer, a substrate of tryptophan synthase, by the racemase reaction and to L-tryptophan by the tryptophan synthase reaction. The industrial procedure for the production of L-tryptophan from DL-serine and indole with both enzymes has been developed by Mitsui-Toatsu Chemicals Inc., Tokyo [68].

3.3.3 Production of L-Lysine from DL-2-Amino-6-caprolactam with L-2-Amino-6-caprolactamase and 2-Amino-6-caprolactam Racemase

Fukumura has established an enzymatic method to produce L-lysine from DL-2-amino-6-caprolactam [69-71]. The process is composed of two enzymatic reactions: the selective hydrolysis of L-2-amino-6-caprolactam to L-lysine, and the racemization of D-2-amino-6-caprolactam (Fig. 7) [71]. The L-2-amino-6-caprolactam-hydrolyzing enzyme, L-2-amino-6-caprolactamase (L-lysine lactamase) (EC 3.5.2.11), is found in the cells of *Cryptococcus laurentii* and some other yeasts [69,70]. The purified enzyme from *C. laurentii* has a molecular weight of approximately 85,000 and is activated by $MnCl_2$ [72]. L-α-Amino-6-caprolactam is the only substrate; the D-isomer and 6-caprolactam are not substrates.

2-Amino-6-caprolactam racemase (EC class 5.1.1.) is found in the cells of *Achromobacter obae* and has been purified from the extract [73,74]. It has a molecular weight of about 50,000 and contains one mole of PLP per mole of enzyme. In addition to

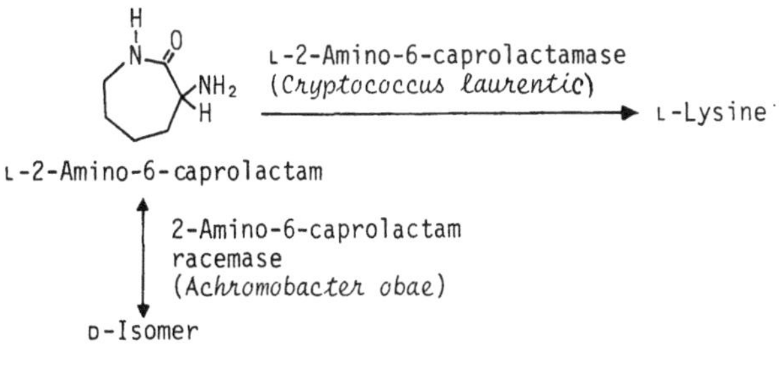

Fig. 7. Enzymatic synthesis of L-lysine from DL-2-amino-6-caprolactam

both the isomers of 2-amino-6-caprolactam, D- and L-2-amino-5-valerolactam also serve as effective substrates.

L-Lysine is industrially synthesized from DL-2-amino-6-caprolactam with a yield of nearly 100% when intact cells of *C. laurentii* and *A. obae* are incubated with the racemate in the same vessel [71]. Recently, Yanai et al. cloned the gene of the racemase of *A. obae* into cells of *E. coli* to produce the enzyme more abundantly [75].

3.3.4 Production of L-Cysteine from DL-2-Amino-Δ^2-thiazolinone-4-carboxylate with 2-Amino-Δ^2-thiazolinone-4-carboxylate Hydrolase, *S*-Carbamoyl-L-cysteine Hydrolase, and 2-Amino-Δ^2-thiazolinone-4-carboxylate Racemase

Sano et al. have isolated several bacterial strains that are capable of producing L-cysteine from DL-2-amino-Δ^2-thiazolinone-4-carboxylate, an intermediate in the chemical synthesis of DL-cysteine [76]. Three enzymes probably participate in this process: L-2-amino-Δ^2-thiazolinone-4-carboxylate hydrolase, *S*-carbamoyl-L-cysteine hydrolase, and 2-amino-Δ^2-thiazolinone-4-carboxylate racemase (Fig. 8). *Pseudomonas thiazolinophilum* isolated from soil has the highest activity of the enzymes that produce L-cysteine from DL-2-amino-Δ^2-thiazolinone-4-carboxylate. Under optimal conditions, 31.4 mg ml^{-1} of L-cysteine was obtained by a mutant of *Ps. thiazolinophilum* lacking L-cysteine-decomposing enzymes such as cysteine desulfhydrase with a molar yield of 95% [77].

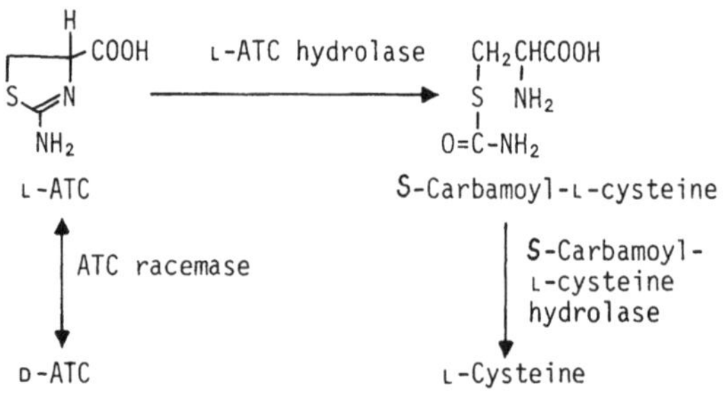

Fig. 8. Enzymatic synthesis of L-cysteine from DL-2-amino-Δ^2-thiazolinone-4-carboxylate (DL-ATC)

3.3.5 Production of D- and L-Lactate from DL-2-Chloropropionate with L-2- and DL-2-Haloacid Dehalogenases

L-2-Haloacid dehalogenase catalyzes dehalogenation of various L-haloalkanoic acids to produce the corresponding 2-hydroxy acids with an inversion of configuration as described above. In contrast, DL-2-haloacid dehalogenase (EC 3.8.1.—), a new enzyme, catalyzes dehalogenation of D- and L-2-haloalkanoic acids [78]. The enzyme purified from *Pseudomonas* sp. 113 has a molecular weight of approximately 68,000 and is composed of two subunits identical in molecular weight (about 35,000) [79]. The substrate specificity of the enzyme is presented in Table 13 [80]. Various DL-2-haloalkanoic acids serve as substrates. The reaction also proceeds in inversion mode. The enzyme works on both the enantiomers of 2-haloacids, but shows stereoselectivity

Table 13. Substrate specificity of DL-2-haloacids dehalogenase

Haloacids	Products	Rel. act.
Monochloroacetic acid	Glycolic acid	33
Monobromoacetic acid	Glycolic acid	280
Monoiodoacetic acid	Glycolic acid	96
Dichloroacetic acid	Glyoxylic acid	8
Trichloroacetic acid	Oxalic acid	3
D-2-Chloropropionic acid	L-Lactic acid	84
L-2-Chloropropionic acid	D-Lactic acid	118
DL-2-Chloropropionic acid	DL-Lactic acid	100
DL-2-Bromopropionic acid	DL-Lactic acid	380
2,2-Dichloropropionic acid	Pyruvic acid	42
DL-2-Chlorobutyric acid	DL-2-Hydroxybutyric acid	18
DL-2-Bromobutyric acid	DL-2-Hydroxybutyric acid	220
DL-2-Bromo-*n*-valeric acid	DL-2-Hydroxyvaleric acid	18

Inert: Monofluoroacetic acid, chloroacetamide, chloroacetaldehyde, 3-chloropropionic acid, 2-chloroisobutyric acid, and 2-chlorocaproic acid

for the substrates; the reaction rate for the D-isomer is about 70% of that for the L-isomers.

The production of D- and L-lactate from DL-chloropropionate has been developed by combination of the reactions of L-2-haloacid dehalogenase and DL-2-haloacid dehalogenase [81]. L-2-Chloropropionate first is converted to D-lactate when L-2-haloacid dehalogenase is incubated with DL-chloropropionate and D-lactate is separated from D-2-chloropropionate in the reaction mixture by ion-exchange chromatography. DL-2-Haloacid dehalogenase, then, is incubated with D-chloropropionate, and consequently L-lactate is formed. The optical purity of D- and L-lactates produced are 98% and 94% with a yield of 90% and 80%, respectively.

4 Enzymatic Synthesis of Optically Active Amino Acids from Achiral Substances

Chiral compounds are used as starting materials in the enzymatic procedures described above, and one enantiomer of the substrate selectively is converted into one enantiomer of the product, or both enantiomers of the substrate are changed to the corresponding enantiomers of the product. In this section the syntheses of optically active amino acids from achiral substances with a single enzyme or multiple enzymes are described.

4.1 Synthesis of L-Amino Acids with Ammonia Lyases

4.1.1 L-Aspartate Production from Fumarate and Ammonia with Aspartase

Aspartase (L-aspartate ammonia lyase) (EC 4.3.1.1) catalyzes the reaction of L-aspartate into fumarate plus ammonia and vice versa. The enzyme occurs widely in plants and microorganisms [82]. The purified enzyme from *Escherichia coli* W has a molecular weight of approximately 193,000 and is composed of four identical

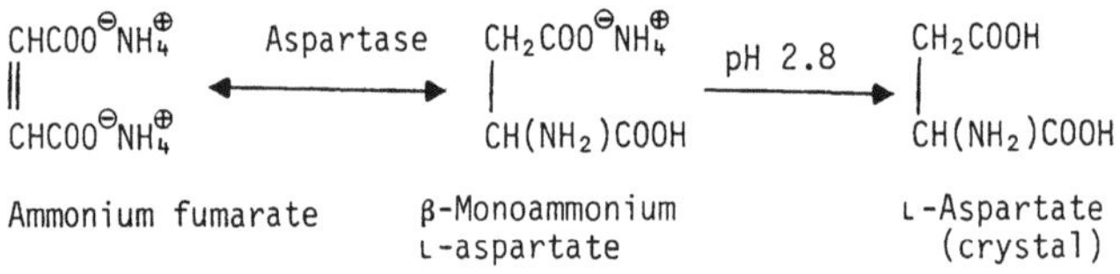

Fig. 9. Enzymatic synthesis of L-aspartate from fumarate and NH_3

subunits (each about 48,000) [83]. Hydroxylamine and hydrazine also serve as substrates alternatively to ammonia. However, D-aspartate, maleate, and mesaconate are inactive. The equilibrium of the reaction favors the amination. In contrast, the equilibrium of phenylalanine ammonia lyase (EC 4.3.1.5) and that of histidine ammonia lyase (EC 4.3.1.3) reactions lies extremely toward deamination [82]. Kitahara et al. have found the highest activity of aspartase in *E. coli* No 2, and developed the procedure for L-aspartate synthesis with the dried cells [84] (Fig. 9). An excess amount of ammonium fumarate is effective to produce ammonium aspartate. L-Aspartate is crystallized by acidification of the reaction mixture. Thus, 56 g of L-aspartate is produced per 100 ml with a molar yield of 99% with 1 g of dried cells. Sato et al. have established a continuous procedure for the industrial production of L-aspartate with *E. coli* cells immobilized in polyacrylamide gel [85].

4.1.2 L-Phenylalanine Production from *trans*-Cinnamate and Ammonia with Phenylalanine Ammonia Lyase

Phenylalanine ammonia lyase (EC 4.3.1.5) catalyzes the conversion of L-phenylalanine to *trans*-cinnamate and ammonia. The enzyme is found in higher plants and microorganisms and has been highly purified and characterized from some of these sources [86]. The equilibrium of the enzyme reaction lies extremely toward deamination of phenylalanine as described above; the amination does not occur in practice under the assay conditions [86]. However, L-phenylalanine is synthesized from *trans*-cinnamate and ammonia when an extremely high concentration of ammonia (e.g., 7.5 M) is added to the reaction system [87] and this can be used for the industrial production of L-phenylalanine, although such a high concentration of ammonia may inactivate the enzyme and make product isolation difficult.

4.2 L-Amino Acid Production from α-Keto Acids and L-Aspartate with Transaminases

L-Amino acids are enzymatically synthesized by a transamination between the corresponding α-keto acids and amino donors. In general, the equilibrium constant for the transamination is about 1.0. As a result, the yield of the desired amino acid for the reaction will never exceed approximately 50%. This problem, incomplete conversion of α-keto acid to L-amino acid, is overcome by using L-aspartate as an amino donor and by converting irreversibly the resulting oxaloacetate to pyruvate with oxaloacetate decarboxylase (EC 4.1.1.3) [88]. Using this method, the enzymatic synthesis of the following L-amino acids in yield approaching 100% have been carried out by employing bacterial transaminases, i.e., phenylalanine, tyrosine, tryptophan, leucine, valine, serine, and methionine.

4.3 L-Serine Production from Glycine with Serine Hydroxymethyltransferase

L-Serine hydroxymethyltransferase (EC 2.1.2.1) catalyzes the cleavage reaction of L-serine to glycine and 5,10-methylenetetrahydrofolate, and the reverse. Although

$$\text{L-Serine} + \text{H}_4\text{-Folate} \rightleftarrows \text{Glycine} + \text{5,10-Methylene-H}_4\text{-folate}$$

tetrahydrofolate, which is not cheaply available, is required *in vivo* for the reaction, the enzymic reaction can be carried out even in the absence of tetrahydrofolate [89]. Therefore, L-serine is enzymatically produced from glycine and formaldehyde.

The enzyme occurs in mammals, insects, and microorganisms, and was purified from a mammalian tissue and characterized enzymologically [89].

4.4 L-Amino Acids Production from α-Keto Acids and Ammonia with Amino Acid Dehydrogenases

Amino acid dehydrogenases catalyze the oxidative deamination of amino acids and the reverse reaction (the reductive amination of α-keto acids) in the presence of $NAD(P)^+$ and NAD(P)H, respectively. L-Glutamate dehydrogenase is ubiquitously found and has been characterized in full detail among various amino acid dehydrogenases [90]. However, the synthesis of L-glutamate with the enzyme may be meaningless in industrial practice because the amino acid is produced efficiently and cheaply by a microbial procedure, and the required starting material, α-ketoglutarate is not as cheap as L-glutamate. The synthesis of L-leucine and L-alanine with bacterial leucine dehydrogenase and alanine dehydrogenase, respectively, are described here.

4.4.1 L-Leucine Production from the α-Keto Analogue and Ammonia with Leucine Dehydrogenase

Leucine dehydrogenase (EC 1.4.1.9) catalyzes the reversible deamination of L-leucine to α-ketoisocaproate and occurs mainly in *Bacillus* species [91]. The enzyme purified and crystallized from *B. sphaericus* has a molecular weight of about 245,000 and consists of six identical subunits (each 41,000) [92]. In addition to branched-chain L-amino acids, straight-chain L-amino acids also are effectively deaminated (Table 14). All the α-keto acids resulting from the oxidative deamination reactions serve as good substrates for the reductive amination reactions (Table 15). The enzyme requires NAD^+ as a coenzyme, whereas NADP is inert.

Wandrey et al. have established an enzymatic procedure for the continuous production of L-leucine from the corresponding keto analogue, α-ketoisocaproate, with a membrane reactor [93]. The membrane reactor involves leucine dehydrogenase, formate dehydrogenase of a yeast, and NADH covalently bound to polyethylene glycol. NAD bound to polyethylene glycol is produced in the reductive amination of α-ketoisocaproate, and the bound NADH is regenerated by the formate dehydrogenase reaction, which proceeds irreversibly in practice to yield carbon dioxide (Fig.

Table 14. Oxidative deamination of amino acids by leucine dehydrogenase from *Bacillus sphaericus*

Amino acids	Rel. act.
L-Leucine	100
L-Valine	74
L-Isoleucine	58
L-Norvaline	41
L-α-Aminobutyrate	14
L-Norleucine	10
γ-Methylallylglycine	8.2
tert-DL-Leucine	1.6
S-Methyl-L-cysteine	1.4
4-Azaleucine	0.6
L-Methionine	0.6
L-Cysteine	0.3
L-Penicillamine	0.2
S-Ethyl-L-cysteine	0.1

Inert: D-Leucine, D-valine, L-alanine, L-glutamate, L-threonine, L-phenylalanine, L-ethionine, L-lysine, and ε-aminocaproate

Table 15. Reductive amination of α-keto acids by leucine dehydrogenase from *Bacillus sphaericus*

α-Keto acids	Rel. act.
α-Ketoisocaproic acid	100
α-Ketoisovaleric acid	126
α-Ketovaleric acid	76
α-Ketobutyric acid	57
α-Ketocaproic acid	46

Inert: pyruvic acid, α-ketoglutaric acid, phenylpyruvic acid, oxaloacetic acid, and glyoxylic acid

10). Thus, α-ketoisocaproate is effectively aminated to L-leucine by coupling both enzyme reactions. Recently, the gene of thermostable leucine dehydrogenase of *B. stearothermophilus* has been cloned in *E. coli* cells with a vector pICR 1 derived from pBR322, and the enzyme was purified from the extracts of the clone cells [94]. The purified enzyme is resistant not only to heating, but also to alkaline and acid treatments and proteolysis by the contaminants.

4.4.2 L-Alanine Production from Pyruvate and Ammonia with Alanine Dehydrogenase

L-Alanine dehydrogenase (EC 1.4.1.1) catalyzes the reversible deamination of L-alanine to pyruvate in the presence of NAD^+, and has been found in a variety of microorganisms. The enzyme was purified to homogeneity and crystallized from *Bacillus sphaericus*, the best producer [95]. It has a molecular weight of approximately

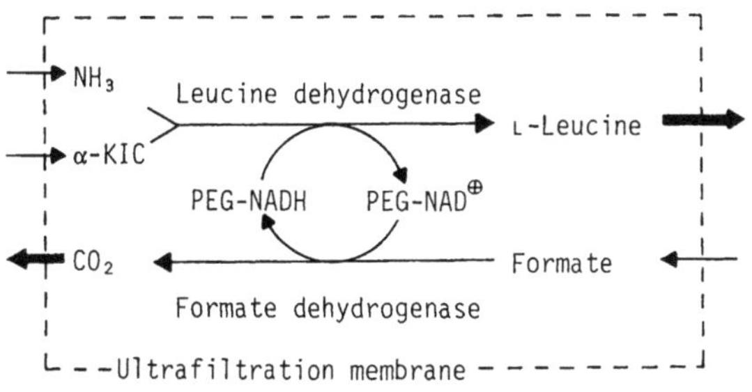

Fig. 10. Enzymatic synthesis of L-leucine with a membrane bioreactor. α-KIC = α-ketoisocaproate. PEG-NAD(H) = polyethyleneglycol-bound NAD(H)

230,000 and is composed of six identical subunits. It acts almost exclusively on L-alanine (Table 16), but shows comparatively low amino acceptor specificity (Table 17). The enzyme is applied to the synthesis of L-alanine from pyruvate with a membrane-bioreactor system as described for leucine dehydrogenase [93].

Table 16. Oxidative deamination of amino acids by alanine dehydrogenase from *Bacillus sphaericus*

Amino acids	Rel. act.
L-Alanine	100
L-α-Aminobutyrate	2.2
L-Serine	1.4
L-Norvaline	0.14
L-Valine	0.07

Inert: D-Alanine, D-serine, glycine, L-leucine, L-glutamate, L-aspartate, L-phenylalanine, L-isoleucine, L-lysine, and β-alanine

Table 17. Reductive amination of α-keto acids by alanine dehydrogenase from *Bacillus sphaericus*

α-Keto acids	Rel. act.
Pyruvic acid	100
α-Ketobutyric acid	103
α-Ketovaleric acid	20.2
β-Hydroxypyruvic acid	3.8
Glyoxylic acid	3.4
α-Ketoisovaleric acid	0.6

Inert: α-ketoglutarate, β-phenylpyruvic acid, and oxaloacetic acid

Sakamoto et al. have cloned the gene of alanine dehydrogenase of *B. stearothermophilus* in *E. coli* C600 with a vector pICR 301 derived from pBR322 [96]. The specific activity of the enzyme was enhanced 50–60 times in the cloned cells, and the enzyme was easily purified from the cell-free extracts by heat-treatment at 70 °C for 30 min. The *E. coli* cells inherently produce D- and L-lactate dehydrogenases abundantly. Based on these findings, Nagata et al. have developed the enzymatic procedure for the synthesis of L-alanine from DL-lactate with the intact cells [97] (Fig. 11). The cells also produce alanine racemase in nature. Thus, the racemase activity was deleted by treatment with a suicide substrate such as D-cycloserine, in order to prevent L-alanine produced from racemization. Alternatively, the gene was cloned in the cells of an alanine racemase-deficient mutant of *E. coli*; the resultant clone cells produce L-alanine from DL-lactate efficiently.

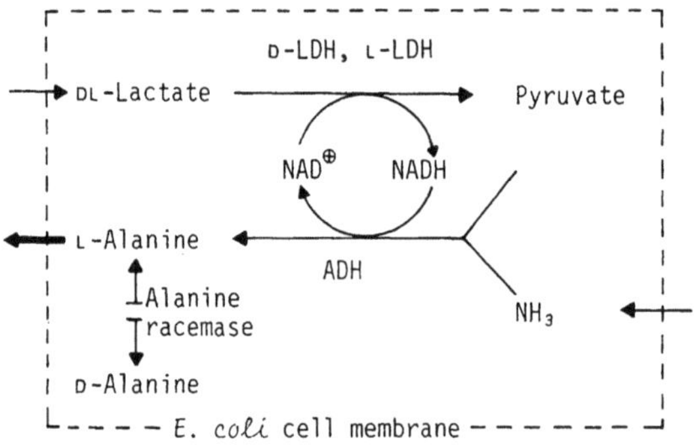

Fig. 11. Enzymatic production of L-leucine with a membrane bioreactor. LDH = lactate dehydrogenase. ADH = L-alanine dehydrogenase

5 Enzymatic Synthesis of Stereospecifically Isotope-labeled Amino Acids, Amines, and Coenzymes

Stereospecifically deuterated or tritiated compounds are useful for the study of the mechanism of various chemical reactions and biological processes. The labeled compounds, however, are not easily prepared by a chemical method. In this Section, enzymatic procedures of stereospecific labeling of amino acids, amines, and coenzymes with deuterium or tritium are described, which are based on the stereoselectivity of the enzymes involved: PLP-dependent enzymes and amino acid dehydrogenases. The stereochemical courses of these enzyme reactions are intensively studied in recent years [5,6].

5.1 Synthesis of Stereospecifically Isotope-labeled Amines and Amino Acids with Amino Acid Decarboxylases

Amino acid decarboxylases catalyze the decarboxylation of amino acids (usually L-isomers) followed by incorporation of one solvent proton to form the corresponding amines. The stereochemistry of this replacement of the carboxyl group by a proton has been investigated with various amino acid decarboxylases [5]. The retention mechanism, in which a proton is introduced to the *pro-R* position of amines, are found for the following PLP-dependent decarboxylases: tyrosine decarboxylase (EC 4.1.1.25) [98], aromatic amino acid decarboxylase (EC 4.1.1.28) [99], lysine decarboxylase (EC 4.1.1.18) [100], glutamate decarboxylase (EC 4.1.1.15) [101], and histidine decarboxylase (EC 4.1.1.22) [102]. Decarboxylation in a retention mode also is found for histidine decarboxylases from *Chlostridium welchii* and *Lactobacillus* 3a which do not require PLP as a cofactor but contain a pyruvoyl residue in the active site [103, 104].

Table 18. Production of stereospecifically tritium-labeled amines and amino acids by amino acid decarboxylases

Enzymes	Substrates	Products
Tyrosine decarboxylase	L-Tyrosine	(1*R*)-(1^3H)-Tyramine
	L-Tryptophan	(1*R*)-(1^3H)-Tryptamine
Aromatic amino acid decarboxylase	α-Methyl-L-DOPA	(1*R*)-(1^3H)-α-Methyl-dopamine
Lysine decarboxylase	L-Lysine	(1*R*)-(1^3H)-Cadaverine
Glutamate decarboxylase	L-Glutamate	(4*R*)-(4^3H)-γ-Aminobutyrate
	(2*S*)-(2^3H)-Glutamate	(4*S*)-(4^3H)-γ-Aminobutyrate
Histidine decarboxylase	L-Histidine	(1*R*)-(1^3H)-Histamine
meso-α,ε-Diaminopimelate decarboxylase	*meso*-α,ε-Diaminopimelate	(6*R*)-(6^3H)-L-Lysine
	(2*R*,6*S*)-*meso*-(2^3H,6^3H)-α,ε-Diaminopimelate	(6*S*)-(6^3H)-L-Lysine
Aspartate β-decarboxylase	(2*S*,3*R*)-(3^3H)-Aspartate	(3*S*)-(3^3H,3^2H)-L-Alanine

In contrast, the decarboxylation with the inversion of configuration occurs in the reactions of *meso*-α,ε-diaminopimelate decarboxylase (EC 4.1.1.26) [105] which releases a carboxyl group from a carbon atom with *R*-configuration, and aspartate β-decarboxylase [106].

When these decarboxylase reactions are carried out in D_2O or T_2O, the amines or the amino acids formed are stereospecifically labeled with deuterium or tritium. For example, (1*R*)-(1^3H)-tyramine, (6*R*)-(6^3H)-L-lysine, and (4*R*)-(4^3H)-γ-aminobutyrate are produced from L-tyrosine, *meso*-α,ε-diaminopimelate, and L-glutamate with the respective decarboxylase reaction in T_2O [98,100,101,107,108]. In addition, the compounds with inversed configuration are prepared by combination of the decarboxylases with the other enzymes; (4*S*)-(4^3H)-γ-aminobutyrate is synthesized by coupling the reaction of glutamate decarboxylase with that of glutamate:pyruvate transaminase [107]. In Table 18 the preparations of such compounds with various amino acid decarboxylases are presented.

5.2 Synthesis of Stereospecifically Isotope-labeled Amino Acids and Pyruvate with Enzymes Catalyzing β-Replacements and α,β-Eliminations

A number of PLP-dependent enzymes catalyze a nucleophilic displacement at C-β of amino acids resulting in the formation of corresponding new amino acids as described above. The steric courses of the reactions of such enzymes have been studied in detail [5]. In the β-replacement reaction both the breaking bond C-β-X and the newly formed bond C-β-Y should be aligned orthogonal to the π plane in the quinoid intermediate (either *syn* or *anti* to the π plane) (Fig. 12). Hence the leaving group X and the incoming group Y can either bind on opposite faces of the π system, leading to reaction with inversion at C-β, or they can bind both on the same face resulting in retention of configuration at C-β. As shown in Table 19, when the enzymes react

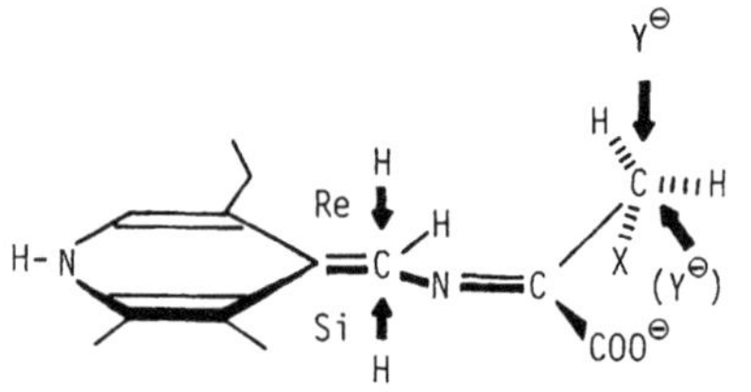

Fig. 12. Structure of the quinoid intermediate

with the stereospecifically labeled substrates at C-β, which are synthesized by an enzymatic method, a chemical method, or by a combination of both methods, the substrate amino acids are converted to the new amino acids labeled stereospecifically at C-β, e.g., (3*S*)-L-cysteine and (3*R*)-L-tyrosine are produced from (3*S*)-*O*-acetyl-L-serine and (3*S*)-L-serine with *O*-acetylserine sulfhydrylase [109] and tyrosine phenol-lyase [111], respectively.

The enzymes functioning in the β-replacement reaction also catalyze the α,β-elimination reaction, resulting in the formation of pyruvate. In the elimination reaction, one proton is stereospecifically introduced into the methyl group of pyruvate. Therefore, when the enzyme reactions are carried out in D_2O (or T_2O), the formed pyruvate is stereospecifically labeled with deuterium (or tritium) at the methyl group; these are listed in Table 20. The configuration of the methyl group produced depends on both the stereoselectivity of the enzyme and the configuration of the substrate.

Table 19. Production of stereospecifically tritium-labeled amino acids by the enzymes catalyzing β-replacement reactions

Enzymes	Substrates	Products
O-Acetylserine sulfhydrylase [109]	(3*S*)-(3^3H)-O-Acetyl-L-serine	(3*S*)-(3^3H)-L-Cysteine
β-Cyanoalanine synthase [110]	(3*S*)-(3^3H)-L-Cysteine	(3*S*)-(3^3H)-L-Cyanoalanine
Tyrosine phenol-lyase [111]	(3*S*)-(3^3H)-L-Serine	(3*R*)-(3^3H)-L-Tyrosine
Tryptophanase [112]	(3*S*)-(3^3H)-L-Serine	(3*R*)-(3^3H)-L-Tryptophan
Tryptophan synthase [113]	(3*S*)-(3^3H)-L-Serine	(3*R*)-(3^3H)-L-Tryptophan

Table 20. Stereospecific labeling of methyl group of pyruvate by the enzymes catalyzing α,β-elimination

Enzymes	Substrates	Products
S-Alkylcysteine lyase [110]	(3*S*)-(3^3H)-L-Cystine	(3*S*)-(3^3H,3^2H)-Pyruvate
	(3*R*)-(3^3H)-L-Cystine	(3*R*)-(3^3H,3^2H)-Pyruvate
Tyrosine phenol-lyase [27]	(3*S*)-(3^3H)-L-Serine	(3*S*)-(3^3H,3^2H)-Pyruvate
	(3*R*)-(3^3H)-L-Serine	(3*R*)-(3^3H,3^2H)-Pyruvate
Tryptophanase [112]	(3*R*)-(3^3H)-L-Tryptophan	(3*S*)-(3^3H,3^2H)-Pyruvate
	(3*S*)-(3^3H)-L-Tryptophan	(3*R*)-(3^3H,3^2H)-Pyruvate
D-Serine dehydratase [114]	(3*S*)-(3^3H)-L-Serine	(3*S*)-(3^3H,3^2H)-Pyruvate
	(3*R*)-(3^3H)-L-Serine	(3*R*)-(3^3H,3^2H)-Pyruvate

5.3 Synthesis of Stereospecifically Isotope-labeled Pyridoxamine with Transaminases

Transaminases catalyze the reversible interconversions of pairs of α-amino and α-keto acids or of terminal primary amines and the corresponding aldehydes by a "shuttle mechanism" in which the enzyme alternates between its PLP form and the corresponding PMP form [115]. In the first half-reaction, the PLP form of the enzyme binds the amino acid (or amine) and forms the coenzyme-substrate Schiff base (I, in Fig. 13). Cleavage of the C-α-H bond is then followed by protonation at C-4′ of PLP (III) through a quinoid intermediate, II as shown in Fig. 13. Hydrolysis of the resulting ketimine then gives a keto acid (or an aldehyde), leaving the enzyme in the PMP form. The steric courses of the protonation have been investigated with various transaminases [5]. The results obtained show that all studied transaminases protonate C-4′ of the PLP-substrate intermediate complex on the *si*-face (Fig. 12), giving *S* configuration at C-4′ of PMP. Thus, stereospecifically deuterium (or tritium-)labeled PMP at C-4′ is produced by carrying out the enzyme reaction in D_2O (or T_2O). The synthesis of stereospecifically labeled PMP with the transaminase reaction, however, may be in practice meaningless, because the coenzyme strongly binds the enzyme protein and is not easily resolved from it. Wada and Snell have demonstrated that various apo-transaminases, including apo-aspertate transaminase (L-aspartate:2-oxoglutarate aminotransferase, EC 2.6.1.1) can reversibly bind pyridoxal (PL) instead of PLP and convert it stoichiometrically into pyridoxamine (PM) in the presence of appropriate L-amino acids [116]. When the reaction with apo-aspartate transaminase is carried out in T_2O, stereospecifically tritiated PM at C-4′ (*S*-configuration) is produced [117]. The same steric course is found for a few other transaminases: alanine transaminase (L-alanine:2-oxoglutarate aminotransferase) (EC 2.6.1.2) [118], pyridoxamine:pyruvate transaminase (EC 2.6.1.30) [119], and dialkylamino acid transaminase [2,2-dialkyl-L-amino acid carboxyl-lyase (aminotransferring)] (EC 4.1.1.64) [120]. Stereospecifically tritium (or deuterium)-labeled PM at C-4′ can be converted to PMP with pyridoxal kinase (EC 2.7.1.35) with retention of the configuration.

5.4 Synthesis of Stereospecifically Isotope-labeled NAD(P)H with Amino Acid Dehydrogenases

It is now generally accepted that all dehydrogenases requiring nicotinamide coenzymes utilize only the 4-position of the nicotinamide ring and that the specificity for 4-*pro-R* or 4-*pro-S* is essentially complete [6]. Amino acid dehydrogenases are not exceptional in this respect. Glutamate dehydrogenase (EC 1.4.1.2-4) reacts with the 4-*pro-S* position of NAD(H) or NADP(H) [121]. The same stereoselectivity is found for the reactions of leucine dehydrogenase [L-leucine:NAD^+ oxidoreductase (deaminating)] (EC 1.4.1.9) [92] and diaminopimelate dehydrogenase [*meso*-2,6-diaminopimelate: $NADP^+$ oxidoreductase (deaminating)] (EC 1.4.1.16) [122]. In contrast, alanine dehydrogenase [L-alanine:NAD^+ oxidoreductase (deaminating)] (EC 1.4.1.1) abstracts a *pro-R* hydrogen [95]. Based on this stereoselectivity, stereospecifically

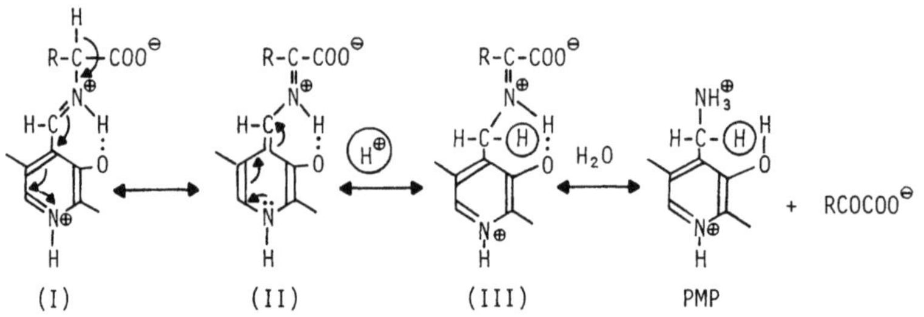

Fig. 13. Conversion of PLP to PMP in the transaminase reaction

tritium- (or deuterium-)labeled NAD(P)H at C-4 is synthesized enzymatically, e.g., (4*S*-^{3}H)NADH (or NADPH) was produced by incubation of bovine liver glutamate dehydrogenase [L-glutamate: $NAD(P)^+$ oxidoreductase (deaminating)] (EC 1.4.1.3) with DL-2-(^{3}H)-glutamate and NAD^+ (or $NADP^+$) [121,92].

5.5 Deuterium- or Tritium-labeling of Amino Acids with Methionine γ-lyase

Methionine γ-lyase catalyzes α,γ-elimination and γ-replacement reactions of L-methionine and its derivatives, and also the α,β-elimination and β-replacement reactions of *S*-substituted L-cysteines as described above. In addition, the enzyme catalyzes the exchange of α- and both β-hydrogens of various amino acids with solvent deuterium or tritium [123]. In the course of the exchange reaction, no racemization occurs, resulting in the formation of the corresponding L-(α-^{2}H, β-$^2H_2$)-amino acids or L-(α-^{3}H, β-$^3H_2$)-amino acids. On the basis of these exchange reactions, Esaki et al. have developed a procedure of enzymatic preparation of α- and β-deuterated or -tritiated amino acids [123]. In addition to L-methionine and S-methyl-L-cysteine, which are good substrates for both the elimination and replacement reactions, straight-chain L-amino acids such as L-alanine and L-norleucine, which are not substrates, undergo the exchange reaction (Table 21).

Table 21. Production of α,β-deuterated or -tritiated amino acids by methionine γ-lyase

Amino acids	Deuterated (%)	Tritiated (dpm per μmol)
L-Methionine	81	8450
S-Methyl-L-cysteine	84	8600
Glycine	—	3900
L-α-Aminobutyrate	92	6940
L-Norleucine	95	7540
L-Norvaline	97	8000
L-Alanine	94	—

Although the rate of α-hydrogen exchange of L-methionine with deuterium is about forty times faster than that of elimination, no hydrogen exchange is found for O-acetylhomoserine, methionine sulfone, and trifluoromethionine which serve as good substrates for α,γ-elimination. None of the following amino acids also are susceptible to the exchange reaction: the D-isomers of the amino acids listed in Table 21 and the L-isomers of α-methyl-methionine, *N*-acetylmethionine, leucine, valine, isoleucine, glutamate, aspartate, glutamine, asparagine, lysine, arginine, and histidine. However, glycine, L-tryptophan, and L-phenylalanine undergo the α-hydrogen exchange, though slowly.

6 Conclusion

Stereospecificity is one of the most prominent properties of enzymes. In addition to the enzymes described here, a variety of other enzymes catalyze stereospecific reactions, e.g., the enzymes involved in the conversion of steroids, coenzyme B_{12}-dependent enzymes and proteases. For example, stereospecific conversions of steroids are performed in organic solvents [124], and the preparation of human insulin from porcine insulin is, in part, based on the stereospecificity of protease [125]. Group rearrangement reactions catalyzed by coenzyme B_{12}-dependent enzymes can be applied for stereospecific synthesis [126]. The preparation of stereospecifically labelled isotopic compounds with these enzymes is also promising. Although studies on asymmetric synthesis have been carried out extensively, chemical production of optically active compounds at present is not feasible in practice. In this respect, enzymes are unsurpassed catalysts. The enzymatic procedures sometimes compete with the chemical ones, but bioindustry essentially should progress in cooperation with the chemical industry, because the former may provide various starting materials for the latter.

Developments of the enzymatic synthesis of optically active compounds also, needless to say, are based on the basic study of the characteristics of enzymes and reaction mechanisms as well as other aspects of biotechnology such as immobilization of biocatalysts and DNA cloning.

7 Abbreviations

PL	pyridoxal
PM	pyridoxamine
PLP	pyridoxal 5′-phosphate
PMP	pyridoxamine 5′-phosphate
NAD	nicotinamide adenine dinucleotide
NADP	nicotinamide adenine dinucleotide phosphate
L-DOPA	3,4-dihydroxy-L-phenylalanine

8 References

1. Wellner, D., Meister, A.: J. Biol. Chem. *235*, 2013 (1960)
2. Dixon, M., Kleppe, K.: Biochim. Biophys. Acta *96*, 368 (1965)
3. Stoops, J. K., Horgan, D. J., Runnegar, T. C., de Jersy, J., Webb, E. C., Zerner, B.: Biochemistry *8*, 2026 (1969)

4. Soda, K., Ohshima, M., Yamamoto, T.: Biochem. Biophys. Res. Commun. *46*, 1278 (1972)
5. Floss, H. G., Vederas, J. C.: Stereochemistry of pyridoxal phosphate-catalyzed reactions, in: New Comprehensive Biochemistry, Vol. 3 (Neuberger, A., van Beenen, L. L. M. eds.), p. 161, Amsterdam: Elsevier Biomedical Press 1982
6. Jeffery, J.: Stereochemistry of dehydrogenases, in: New Comprehensive Biochemistry, Vol. 3 (Neuberger, A., van Beenen, L. L. M. eds.), p. 113, Amsterdam: Elsevier Biomedical Press 1982
7. Breslow, R., Hammond, M., Lauer, M.: J. Am. Chem. Soc. *102*, 421 (1980)
8. Soda, K., Tanaka, H., Esaki, N.: Amino Acids, in: Biotechnology, Vol. 3 (Rehm, H. J., Reed, G. eds.), p. 479, Weinheim: Verlag Chemie 1983
9. Chemical and Biological Aspects of Vitamin B_6 Catalysis, Part A and B (Evangelopolous, A. E., ed.), New York, Alan R. Liss, Inc. 1984
10. Tanaka, H., Esaki, N., Soda, K.: Biochemistry *16*, 100 (1977)
11. Tanaka, H., Esaki, N., Yamamoto, T., Soda, K.: FEBS Lett. *66*, 307 (1976)
12. Davis, L., Metzler, D. E.: Pyridoxal-linked elimination and replacement reactions, in: The Enzymes, Vol. 7 (Boyer, P. D. ed), p. 33, New York: Academic Press 1972
13. Esaki, N., Suzuki, T., Tanaka, H., Soda, K., Rando, R. R.: FEBS Lett. *84*, 309 (1977)
14. Esaki, N., Tanaka, H., Uemura, S., Suzuki, T., Soda, K.: Biochemistry *18*, 407 (1979)
15. Kumagai, H., Yamada, H., Matsui, H., Ohkishi, H., Ogata, K.: J. Biol. Chem. *245*, 1767 (1970)
16. Kumagai, H., Kashima, N., Yamada, H., Enei, H., Okumura, S.: Agric. Biol. Chem. *36*, 472 (1972)
17. Kumagai, H., Ohkishi, H., Kashima, N., Yamada, H.: Amino Acids Nucleic Acids *27*, 11 (1973)
18. Ueno, T., Fukami, H., Ohkishi, H., Kumagai, H., Yamada, H.: Biochim. Biophys. Acta *206*, 476 (1970)
19. Enei, H., Matsui, H., Nakazawa, H., Okumura, S., Yamada, H.: Agric. Biol. Chem. *37*, 493 (1973)
20. Enei, H., Nakazawa, H., Okumura, S., Yamada, H.: ibid. *37*, 727 (1973)
21. Ikeda, S., Fukui, S.: Methods Enzymol. *62*, 517 (1979)
22. Yamada, H., Yamada, K., Kumagai, H., Hino, T., Okumura, S.: Immobilization of β-tyrosinase cells with collagen, in: Enzyme Engineering, Vol. 3 (Pye K., Weetall, H. H. eds.), p. 57, New York: Plenum Press 1978
23. Yamada, H., Kumagai, H.: Adv. Appl. Microbiol. *19*, 249 (1975)
24. Kumagai, H., Yamada, H., Matsui, H., Ohkishi, H., Ogata, K.: J. Biol. Chem. *245*, 1773 (1970)
25. Muro, T., Nakatani, H., Hiromi, K., Kumagai, H., Yamada, H.: J. Biochem. *84*, 633 (1978)
26. Sawada, S., Kumagai, H., Yamada, H., Hill, R. K.: J. Am. Chem. Soc. *97*, 4334 (1975)
27. Kumagai, H., Yamda, H., Sawada, S., Schleider, E., Mascaro, K., Floss, H. G.: J. Chem. Soc. Chem. Commun. 85 (1977)
28. Miles, E. W.: Adv. Enzymol. *49*, 127 (1979)
29. Esaki, N., Tanaka, H., Miles, E. W., Soda, K.: Agric. Biol. Chem. *47*, 2861 (1983)
30. Esaki, N., Tanaka, H., Miles, E. W., Soda, K.: FEBS Lett. *161*, 207 (1983)
31. Newton, W., Snell, E. E.: Proc. Natl. Acad. Sci. U.S.A. *51*, 381 (1964)
32. Yoshida, H., Kumagai, H., Yamada, H.: Agric. Biol. Chem. *38*, 463 (1974)
33. Newton, W., Morino, H., Snell, E. E.: J. Biol. Chem. *240*, 1211 (1965)
34. Yamagata, S., Takeshima, K.: J. Biochem. *80*, 777 (1976)
35. Chocat, P., Esaki, N., Tanaka, H., Soda, K.: Agric. Biol. Chem. *49*, 1143 (1985)
36. Esaki, N., Nakamura, T., Tanaka, H., Suzuki, T., Morino, Y., Soda, K.: Biochemistry *20*, 4492 (1981)
37. Nagasawa, T., Ohkishi, H., Kawakami, B., Yamano, H., Hosono, H., Tani, Y., Yamada, H.: J. Biol. Chem. *257*, 13749 (1982)
38. Yamada, H., Nagasawa, T., Ohkishi, H., Kawakami, B., Tani, Y.: Biochem. Biophys. Res. Commun. *100*, 1104 (1981)
39. Nagasawa, T., Yamano, H., Ohkishi, H., Tani, Y., Yamada, H.: Agric. Biol. Chem. *46*, 3003 (1982)
40. Wilson, E. W., Kornberg, H. L.: Biochem. J. *88*, 578 (1963)
41. Kakimoto, T., Kato, J., Shibatani, T., Nishimura, N., Chibata, I.: J. Biol. Chem. *244*, 353 (1969)
42. Chibata, I., Kakimoto, T., Kato, J.: Appl. Microbiol. *13*, 638 (1965)
43. Motosugi, K., Esaki, N., Soda, K.: Agric. Biol. Chem. *46*, 837 (1982)
44. Chibata, I.: Optical resolution of DL-amino acids, in: Synthetic Production and Utilization of

Amino Acids (Kaneko, T., Izumi, Y., Chibata, I., Itoh, T., eds.), p. 17, Tokyo: Kodansha Ltd. and New York: Wiley & Sons 1974
45. Gil-Ave, E., Tishbee, A., Hare, P. E.: J. Am. Chem. Soc. *102*, 5116 (1980)
46. Davankov, V. A., Semechkin, A. V.: J. Chromatogr. *141*, 313 (1977)
47. Manning, J. M., Moore, S.: J. Biol. Chem. *243*, 559 (1968)
48. Doherty, D. G., Popenoe Jr., E. A.: ibid. *189*, 447 (1951)
49. Albertson, N. F.: J. Am. Chem. Soc. *73*, 452 (1951)
50. Greenstein, J. P.: Methods Enzymol. *3*, 554 (1957)
51. Rao, K. R., Birnbaum, R. B., Kingsley, R. B., Greenstein, J. P.: J. Biol. Chem. *198*, 507 (1952)
52. Chibata, I., Ishikawa, T., Yamada, S.: Bull. Agric. Chem. Soc. Jpn. *21*, 300 (1957)
53. Tosa, T., Mori, T., Fuse, N., Chibata, I.: Enzymologia *31*, 214 (1966)
54. Paik, W. K., Block-Frankenthal, L., Birnbaum, S. M., Winitz, M., Greenstein, J. P.: Arch. Biochem. Biophys. *69*, 56 (1957)
55. Kameda, T., Toyoura, E., Kimura, Y., Marsui, K.: Chem. Pharm. Bull. (Tokyo) *6*, 394 (1958)
56. Chibata, I., Ishikawa, T., Tosa, T.: Methods Enzymol. *19*, 756 (1970)
57. Bucherer, H. T., Steiner, W.: J. Prakt. Chem. *140*, 291 (1934)
58. Wallach, D. P., Grisolia, S.: J. Biol. Chem. *226*, 277 (1957)
59. Yamada, H., Takahashi, S., Kii, Y., Kumagai, H.: J. Ferment. Technol. *56*, 484 (1978)
60. Takahashi, S., Kii, Y., Kumagai, H., Yamada, H.: ibid. *56*, 492 (1978)
61. Yamada, H., Shimizu, S., Shimada, H., Tani, Y., Takahashi, S., Ohashi, T.: Biochemie *62*, 395 (1980)
62. Sano, K., Yokozeki, K., Eguchi, C., Kagawa, T., Noda, I., Mitsugi, K.: Agric. Biol. Chem. *41*, 819 (1977)
63. Schwert, G. W., Neurath, H., Kaufman, S., Snoke, J. E.: J. Biol. Chem. *172*, 221 (1948)
64. Greenstein, J. P., Winitz, M.: Chemistry of the Amino Acids, p. 715, New York—London: Wiley & Sons 1961
65. Soda, K., Yonaha, K., Misono, H., Osugi, M.: FEBS Lett. *46*, 359 (1974)
66. Tanaka, H., Yamada, N., Esaki, N., Soda, K.: Agric. Biol. Chem. *49*, 2525 (1985)
67. Soda, K., Osumi, T.: Methods Enzymol. *17B*, 629 (1971)
68. Makiguchi, N., Fukuhara, N., Shimada, M., Asai, Y., Soda, K.: Abstracts of Ann. Meeting of Agric. Chem. Soc. Japan, p. 344, 1985
69. Fukumura, T.: Agric. Biol. Chem. *40*, 1687 (1976)
70. Fukumura, T.: ibid. *40*, 1695 (1976)
71. Fukumura, T.: ibid. *41*, 1327 (1977)
72. Fukumura, T., Talbot, G., Misono, H., Teramura, Y., Kato, K., Soda, K.: FEBS Lett. *89*, 298 (1978)
73. Fukumura, T.: Agric. Biol. Chem. *41*, 1321 (1977)
74. Ahmed, S. A., Esaki, N., Tanaka, H., Soda, K.: ibid. *47*, 1149 (1983)
75. Yanai, A., Nakamura, N., Oshihara, W.: Abstracts of Ann. Meeting of Agric. Chem. Soc. Japan, p. 422, 1985
76. Sano, K., Yokozeki, K., Tamura, F., Yasuda, N., Noda, I., Mitsugi, K.: Appl. Environ. Microbiol. *34*, 806 (1977)
77. Sano, K., Matsuda, K., Yasuda, N., Mitsugi, K.: Amino Acids Nucleic Acids *35*, 112 (1977)
78. Motosugi, K., Esaki, N., Soda, K.: Arch. Microbiol. *131*, 179 (1982)
79. Motosugi, K., Esaki, N., Soda, K.: J. Bacteriol. *150*, 522 (1982)
80. Motosugi, K., Soda, K.: Experientia *39*, 1214 (1983)
81. Motosugi, K., Esaki, N., Soda, K.: Biotech. Bioeng. *26*, 805 (1984)
82. Hanson, J. R., Havir, E. A.: The enzymatic elimination of ammonia, in: The Enzymes Vol. 7 (Boyer, P. D., ed.), p. 75, New York: Academic Press, 1972
83. Suzuki, S., Yamaguchi, J., Tokushige, M.: Biochim. Biophys. Acta *321*, 369 (1973)
84. Kitahara, K., Fukui, S., Misawa, M.: Nippon Nogeikagaku Kaishi *34*, 44 (1960)
85. Sato, T., Mori, T., Tosa, T., Chibata, I., Furui, M., Yamashita, K., Sumi, A.: Biotech. Bioeng. *17*, 1797 (1975)
86. Ogata, K., Uchiyama, K., Yamada, H., Tochikura, T.: Agric. Biol. Chem. *31*, 600 (1967)
87. Yamada, S., Nobe, K., Izuo, N., Nakamachi, K., Chibata, I.: Appl. Environ. Microbiol. *42*, 773 (1981)
88. Rozzell, J. D.: U.S. Patent Number 4,518,693 (May 21, 1985)

89. Schirch, L.: Adv. Enzymol. *53*, 83 (1982)
90. Smith, E. L., Austen, B. M., Blumenthal, K. M., Nyc, J. F.: Glutamate dehydrogenases, in: The Enzymes Vol. 9 (Boyer, P. D., ed.), p. 294, New York: Academic Press 1975
91. Soda, K., Misono, H., Mori, K., Sakato, H.: Biochem. Biophys. Res. Commun. *44*, 931 (1971)
92. Ohshima, T., Misono, H., Soda, K.: J. Biol. Chem. *253*, 5719 (1978)
93. Wandrey, C., Wichman, R., Jandel, A.-S.: Multienzyme systems in membrane reactor, in: Enzyme Engineering Vol. 6 (Chibata, I., Fukui, S., Wingrad Jr., L. B., eds.), p. 61, New York: Plenum Press 1982
94. Nagata, S., Sakamoto, Y., Esaki, N., Ohshima, T., Tanaka, H., Soda, K.: Abstracts of Ann. Meeting of Agric. Chem. Soc. Japan, p. 3, 1983
95. Ohshima, T., Soda, K.: Eur. J. Biochem. *100*, 29 (1979)
96. Sakamoto, Y., Nagata, S., Inagaki, K., Ohshima, T., Tanaka, H., Soda, K.: Abstracts of Ann. Meeting of Agric. Chem. Soc. of Japan, p. 113, 1984
97. Nagata, S., Esaki, N., Tanaka, H., Soda, K.: Abstracts of Ann. Meeting of Agric. Chem. Soc. Japan, p. 594, 1984
98. Battersby, A. R., Nicholetti, M., Staunton, J., Vleggaar, R.: J. Chem. Soc. Perkin Trans. *1*, 43 (1980)
99. Marshall, K. S., Castagnoli Jr., N.: J. Med. Chem. *16*, 266 (1973)
100. Gerdes, H. J., Leistner, E.: Phytochemistry *18*, 77 (1979)
101. Yamada, H., O'Leary, M. H.: Biochemistry *17*, 669 (1978)
102. Battersby, A. R., Joyeau, R., Staunton, J.: FEBS Lett. *107*, 231 (1979)
103. Battersby, A. R., Chrystal, E. J. T., Staunton, J.: J. Chem. Soc. Perkin Trans. *1*, 31 (1980)
104. Santaniello, E., Manzocchi, A., Biondi, P. A.: ibid. *1*, 307 (1981)
105. Asada, Y., Tanizawa, K., Sawada, S., Suzuki, T., Misono, H., Soda, K.: Biochemistry *20*, 6881 (1981)
106. Chang, C.-C., Laghi, A., O'Leary, M. H., Floss, H. G.: J. Biol. Chem. *257*, 3564 (1982)
107. Burnett, G., Walsh, C., Yonaha, K., Toyama, S., Soda, K.: J. Chem. Soc. Chem. Commun. 826 (1979)
108. Tanizawa, K., Yoshimura, T., Asada, Y., Sawada, S., Misono, H., Soda, K.: Biochemistry *21*, 1104 (1982)
109. Floss, H. G., Schlicher, E., Rotts, R.: J. Biol. Chem. *251*, 5478 (1976)
110. Tsai, M. D., Weaver, J., Floss, H. G., Conn, E. E., Creaveling, R. K., Mazelis, M.: Arch. Biochem. Biophys. *190*, 553 (1978)
111. Fuganti, C., Ghiringhelli, D., Giangrasso, D., Grasselli, P.: J. Chem. Soc. Chem. Commun. 726 (1974)
112. Vederas, J. C., Schlicher, E., Tsai, M. D., Floss, H. G.: J. Biol. Chem. *253*, 3350 (1978)
113. Tsai, M. D., Schlicher, E., Potss, R., Skye, G. E., Floss, H. G.: ibid. *253*, 5344 (1978)
114. Cheung, Y. F., Walsh, C.: J. Am. Chem. Soc. *98*, 3397 (1976)
115. Braunstein, A. E.: Amino group transfer, in: The Enzymes Vol. 9 (Boyer, P. D., ed.), p. 379, New York: Academic Press 1973
116. Wada, H., Snell, E. E.: J. Biol. Chem. *237*, 127 (1962)
117. Dunathan, H. C., Davis, L., Gilmer Kury, P., Kaplan, M.: Biochemistry *7*, 4532 (1968)
118. Austermühle-Bertola, E.: Ph.D. Dissertation No. 5009, ETH, Zurich 1973
119. Ayling, J. E., Dunathan, H. C., Snell, E. E.: Biochemistry, *7*, 4537 (1968)
120. Bailey, G. B., Kusamrarn, T., Vuttivej, K.: Fed. Proc. *29*, 857 (1970)
121. Levy, H. R., Vennesland, B.: J. Biol. Chem. *228*, 85 (1957)
122. Misono, H., Soda, K.: ibid. *255*, 10599 (1980)
123. Esaki, N., Sawada, S., Tanaka, H., Soda, K.: Anal. Biochem. *119*, 281 (1982)
124. Fukui, S., Tanaka, A.: Advances in Biochemical Engineering/Biotechnology *29*, 1 (1984)
125. Morihara, K., Oka, T., Tsuzuki, H.: Nature *280*, 412 (1979)
126. Rétey, J.: Vitamin B_{12}: Stereochemical aspects of its biological functions and of its biosynthesis, in: New Comprehensive Biochemistry Vol. 3 (Neuberger, A., van Deenen, L. L. M., eds.), p. 249, Amsterdam: Elsevier Biomedical Press 1982

Fluidized Bed Biofilm Reactor for Wastewater Treatment

Wen K. Shieh and John D. Keenan
Department of Civil Engineering,
University of Pennsylvania,
Philadelphia, Pennsylvania 19104 U.S.A.

1 Introduction ... 132
1.1 Process Description of Fluidized Bed Biofilm Reactor ... 132
1.2 Comparison of Fluidized Bed Biofilm Reactor with Competing Biological Wastewater Treatment Processes ... 133
2 Microbiology of Fluidized Bed Biofilm Reactor ... 134
2.1 Occurrence of Biofilms on Inert Media ... 134
2.2 Biofilm Characteristics ... 135
2.2.1 Properties of Biofilms ... 136
2.2.2 Physical Properties of Biofilms ... 136
2.2.3 Chemical Properties of Biofilms ... 138
2.2.4 Biological Properties of Biofilms ... 140
2.3 Factors Affecting Biofilm Formation on the Inert Media ... 141
2.3.1 Polysaccharide Materials ... 141
2.3.2 Media Surface Characteristics ... 141
2.3.3 Microbial Species ... 141
2.3.4 Hydrodynamic Environment ... 141
2.3.5 Substrate Characteristics and Concentration ... 142
2.3.6 Environmental Conditions ... 142
3 Process Kinetics of Fluidized Bed Biofilm Reactor ... 143
3.1 Mass Transfer Limitations – Concept of Effectiveness Factor ... 143
3.1.1 External Mass Transfer ... 144
3.1.2 Effectiveness Factor Expressions ... 145
3.2 Fluidization Mechanics ... 149
3.3 Overall Rates of Substrate Conversion ... 151
3.3.1 Plug-Flow Conditions (Recycle Ratio $\leqq 2$) ... 152
3.3.2 Complete-Mix Conditions (Recycle Ratio >2) ... 153
3.4 Limiting Bulk-Liquid Substrate Concentration ... 153
3.5 Critical Biofilm Thickness ... 154
3.6 Critical Media Size and Density ... 157
3.7 Parameter Estimation ... 159
3.7.1 Zero Order Kinetics ... 159
3.7.2 First Order Kinetics ... 159
3.7.3 Fluidization Correlations ... 159
4 Design Considerations of Fluidized Bed Biofilm Reactor ... 160
4.1 Distribution of Incoming Wastewater ... 160
4.2 Minimum Fluidization Velocity ... 161
4.3 Control of Expanded Bed Height ... 161
4.4 Biofilm Separation and Biomass Wastage ... 162
4.5 Recirculation Requirements ... 163
4.6 Addition of Oxygen and Chemicals ... 163
5 Current Applications ... 164
6 Future Outlook ... 165
7 Notation ... 166
8 References ... 167

The fluidized bed biofilm reactor (FBBR) represents a recent innovation in biofilm processes. Immobilization of microorganisms on the small, fluidized particles of the medium results in a high reactor biomass holdup which enables the process to be operated at significantly higher liquid throughputs with the practical absence of biomass wash-out. The process intensification (i.e., a reduction in process size while maintaining performance) achieved in FBBRs makes this innovative technology particularly attractive in biological wastewater treatment, commercial biomass conversion, and ethanol and biochemical production applications. In this chapter, the present understanding of biofilm phenomena involved in the operation of FBBRs is reviewed. Special emphasis is placed on the microbial and kinetic aspects of FBBRs and practical design considerations and current applications are described.

1 Introduction

A solid surface in contact with nutrient medium containing microorganisms will eventually be covered with biofilms because of the adhesion of microorganisms from the bulk liquid. This phenomenon forms the cornerstone of some industrially important processes which utilize biofilms. Prominent examples include the trickling filter process in wastewater treatment, the "quick" vinegar process [1] and bacterial leaching [2].

Growth support media (referred hereafter to as media) within conventional biofilm reactors are fixed in space either by gravity or by direct attachment to the reactor wall. In contrast, the reactor to be discussed in this chapter, which represents a recent innovation in biofilm processes, retains media in suspension by drag forces exerted by the upflowing nutrient medium. This new type of reactor is referred to as a fluidized bed biofilm reactor (FBBR).

1.1 Process Description of Fluidized Bed Biofilm Reactor

A schematic diagram of an FBBR for wastewater treatment is illustrated in Fig. 1. Wastewater to be treated is pumped upward through a bed of media (e.g., silica sand) at velocities sufficient to induce fluidization of the media. Once fluidized, each particle provides a large surface area for biofilm formation and growth. As the biofilms

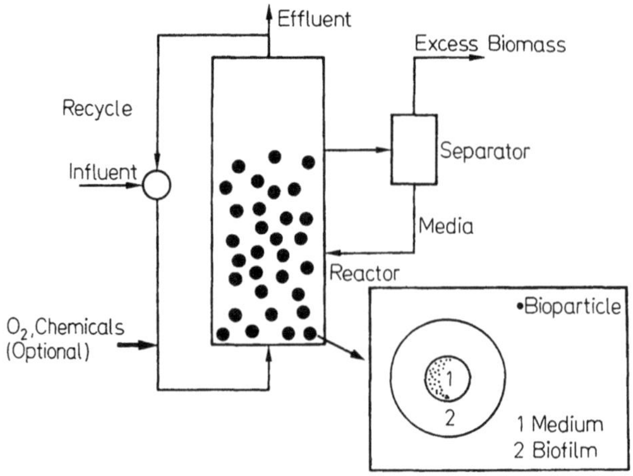

Fig. 1. Schematic diagram of a fluidized bed biofilm reactor

cover the media, the overall density of the biofilm-coated media (referred hereafter to as bioparticles) decreases, which eventually would cause the wash-out of the bioparticles from the reactor. This can be prevented by controlling the expanded bed height at a given level via intentional wasting of the overgrown bioparticles. Usually a mechanical device is used to separate the biomass from the wasted bioparticles. The cleaned media are returned to the reactor whereas the separated biomass is wasted as the excess sludge. In most cases, recycle of reactor effluent is employed to insure uniform fluidization and adequate substrate loading rate.

1.2 Comparison of Fluidized Bed Biofilm Reactor with Competing Biological Wastewater Treatment Processes

A comparison of FBBR with competing biological wastewater treatment processes in municipal applications, in terms of specific surface area, biomass concentration and process loading rate, is summarized in Table 1 [3].

The use of small, fluidized media enables the FBBR to retain high biomass concentrations and thereby operate at significantly reduced hydraulic retention times. Jeris and Owen [4] and Jeris, et al. [5] have reported volatile solids concentrations between 30,000 and 40,000 mg L^{-1} for pilot-scale denitrification studies employing FBBRs. As a result, 99% of influent nitrates can be removed at empty bed hydraulic retention times as low as 6 minutes.

Fluidization also overcomes operating problems such as bed clogging and the high pressure drop which would occur if small, high surface area media were employed in packed bed operation. A further advantage is the possible elimination of the secondary clarifier, although this must be weighed against the medium-biomass separator.

Table 1. Comparison of FBBR with competing biological wastewater treatment processes in municipal applications [3]

Treatment process	Parameter
	(1) Specific surface area of media (m^2 m^{-3} reactor volume)
• Trickling filter	12–30
• Rotating biological contactor	40–50
• FBBR	800–1,200
	(2) Biomass concentration (mg L^{-1})[a]
• Pure oxygen activated sludge	3,000–5,000
• Conventional activated sludge	2,000–3,000
• Nitrification activated sludge	1,000–1,500
• FBBR (Carbon oxidation)	12,000–15,000
• FBBR (Nitrification)	8,000–12,000
• FBBR (Denitrification)	30,000–40,000
	(3) Process loading rate[b]
• Pure oxygen activated sludge	1.2–2.4
• Conventional activated sludge	0.5–1.2
• FBBR (Carbon oxidation)	8–16

[a] In terms of mixed liquor volatile suspended solids (MLVSS) concentration
[b] In terms of kg BOD removed per m^3 reactor volume per day

FBBRs have been investigated, at least to pilot scale, for all of the basic secondary and tertiary treatment processes, including carbon oxidation [6,7], nitrification [8], denitrification [9-13], and anaerobic treatment [14-19], for a variety of waters and municipal and industrial wastewaters.

2 Microbiology of Fluidized Bed Biofilm Reactor

Because all the claimed advantages of FBBRs are derived, without exception, from their ability of retain high biomass concentrations as biofilms within the reactors, an understanding of the phenomenon of biofilm formation is essential to the successful application of the FBBR technology. Unfortunately, there is a paucity of information on the dynamics of biofilm formation on fluidized media. Nevertheless, research work done by marine biologists, microbiologists and engineers in other disciplines has provided a good amount of data of great value in the understanding of the FBBR microbiology [20,21].

2.1 Occurrence of Biofilms on Inert Media

Zobell and coworkers [22-24] were among the first researchers to observe and report that the presence of solid surfaces in seawater samples can enhance microbial activity. Zobell [24] further demonstrated that the formation of biofilm on the solid surface is required for microbial growth at low substrate concentration. Similar phenomena were observed by Heukelekian and Heller [25], Heukelekian and Crosby [26], Jannasch [27], and Bott and Miller [28].

Biofilm development can be represented in a fashion as illustrated in Fig. 2 [29]. In essence, biofilm development on an inert surface occurs in five distinct phases [29-32]. Initially, microorganisms suspended in the nutrient medium are transported to the support medium surface via different mechanisms such as Brownian motion, electrostatic attraction, van der Waals force, turbulent eddy transport, sedimentation and thermophoresis [29-32]. Once attached, microorganisms require time to acclimate to their new environment before they utilize the substrates already adsorbed on the media surface. Thus, there is a lag phase between the initial attachment of microorganisms and the subsequent accumulation of biofilms on the media surface. The second phase is logarithmic in which microorganisms grow at a rate determined by their generation time and their ability to process substrate. This phase will last until the whole surface is covered with biofilm and the "active" biofilm thickness is reached. At this point, the increase in mass of microorganisms shifts from logarithmic to linear with time. During the third phase, biofilm growth is essentially linear and biofilm continues to accumulate until the "plateau" biofilm thickness is reached. This is the beginning of the fourth phase in which the microorganisms within the biofilm have exhausted the substrate surrounding them and the growth of new cells is offset by the death of old ones. Portions of biofilm are detached from the media surface and entrained in the fluid flow. The sloughing phase represents the period in which the rate of biofilm detachment is substantial which results in a net decrease in biofilm mass and thickness.

Sutton et al. [8] have discussed the accumulation of biofilms on the fluidized media during the start-up of an FBBR designed for combined carbon oxidation-nitrification

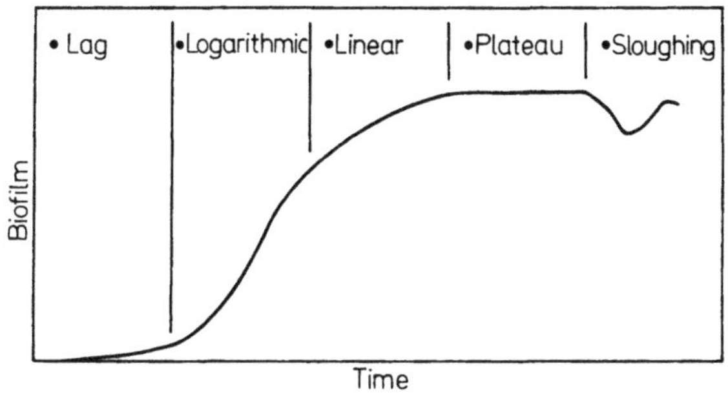

Fig. 2. Biofilm development on the surface of inert media [29]

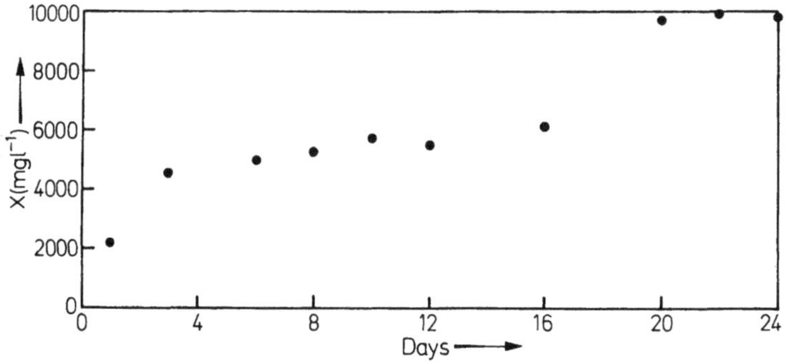

Fig. 3. Accumulation of biofilm on the fluidized media (in terms of biomass concentration, X) during the start-up of an FBBR for combined carbon oxidation-nitrification of a municipal wastewater [8]. Silica sand was used as the media

of a municipal wastewater. Mean influent BOD_5 and TKN were 67 and 23 mg L^{-1}, respectively. Silica sand with a mean diameter of approximately 0.5 mm was used as the medium. A hydraulic loading rate of 36.6 m h^{-1} and a recycle ratio of 1.4 were employed for start-up which resulted in an empty bed hydraulic retention time of approximately 8 minutes. Their results, in terms of mean FBBR biomass concentrations, are illustrated in Fig. 3. Although the data shown in Fig. 3 are insufficient to exhibit the clear distinction among different growth phases as illustrated in Fig. 2, it appears that biofilm accumulation reaches the "plateau" phase after about three weeks of operation.

2.2 Biofilm Characteristics

Biofilm has been defined as the gelatinous matrix formed on media surfaces and is primarily composed of carbohydrates [33] which are polymerized to produce branched or linear polysaccharides [34–35]. The biofilm acts as a binder which holds large colonies of microorganisms together [36]. The polysaccharide materials in biofilms are considered to be microbial non-viable excretions or secretions. These materials tend to compact in areas adjacent to the microorganisms [21].

2.2.1 Properties of Biofilms

Physical, chemical and biological properties of biofilms are dependent on the environment to which the media surface is exposed. Moreover, as the biofilms accumulate, changes in biofilm properties may occur.

2.2.2 Physical Properties of Biofilms

Biofilm Thickness

Although the total mass of biofilms in biofilm reactors is a direct function of biofilm thickness, it is important to realize that a thick biofilm does not necessarily have a greater substrate conversion rate than a thin biofilm because of the mass transfer resistances imposed by the gelatinous structure of the biofilm. This conclusion is substantiated by the experimental evidence reported by Mulcahy, et al. [37] which is illustrated in Fig. 4. As shown, the observed denitrification rate in a rotating-disk biofilm reactor increases linearly with biofilm thickness up to a certain thickness, beyond which it remains constant. When the biofilm thickness is less than a critical value, the whole biofilm is active. As the biofilm thickness exceeds the critical value, only that portion of the biofilm with thickness less than the critical value is still active. As a result, further increase in biofilm thickness will not induce a corresponding increase in the observed substrate conversion rate.

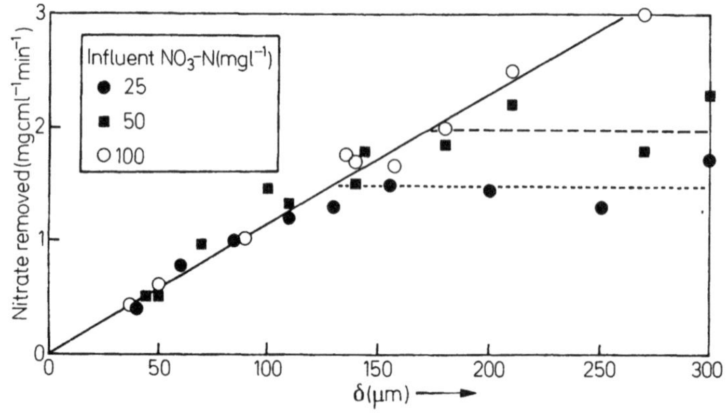

Fig. 4. The observed denitrification rate in a rotating-disk biofilm reactor as a function of biofilm thickness (δ) [37]

Atkinson and Fowler [38] reported that the hydrodynamic characteristics of biofilm reactors have profound effects on equilibrium biofilm thickness. Moreover, Characklis [39] demonstrated that, in an annular reactor, the equilibrium biofilm thickness attained at a given fluid shear stress is a direct function of the substrate loading rate (see Fig. 5).

The situation is somewhat different in an FBBR because the biofilm thickness can be maintained at a relatively constant value by controlling the fluidized bed height and by wasting excess biomass regularly. Shieh, et al. [40] reported that the biofilm thickness could be maintained at approximately 150 microns in an FBBR designed for

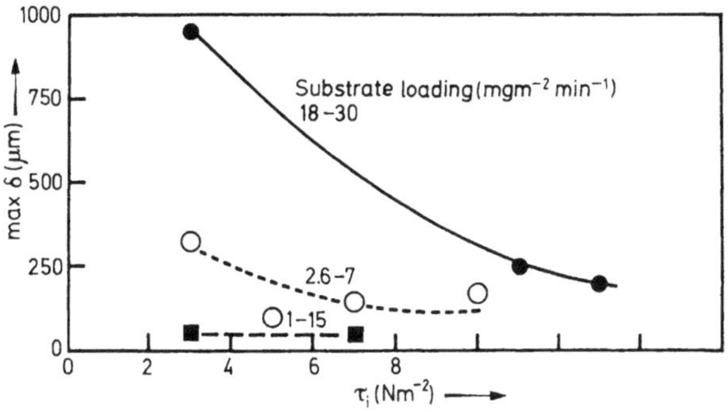

Fig. 5. Effects of fluid shear stress and substrate loading rate on the equilibrium biofilm thickness [39]. τ_i: initial fluid shear stress

carbon oxidation of a concentrated corn starch wastewater (influent BOD_5 = 2,000 mg L^{-1}) via this strategy in spite of the fact that the substrate loading rates were varied over a wide range.

Biofilm Dry Density

Because the substrate conversion rates in biofilm reactors are a function of biomass, these rates may be calculated using both biofilm dry density and thickness [41]. Hoehn and Ray [42] reported that the biofilm dry density of biofilm varies with biofilm thickness and reaches a maximum value at a thickness consistent with the active biofilm thickness. They postulated that the changes in biofilm dry density are caused by the variations in the physiological state of the microorganisms. When the biofilm is thin, and/or the bulk-liquid substrate concentration is high, microbial growth is unimpeded by a lack of substrate because of full penetration of substrate within the biofilm. This results in a higher biofilm dry density. As the biofilm thickness grows, or the bulk-liquid substrate concentration decreases, a transition occurs that causes substrate utilization, and therefore microbial growth, within the biofilm to become mass transfer-limited. In this case, the microorganisms in the inner part of the biofilm will enter the endogeneous respiration phase, and cause a decrease in biofilm dry density.

Mulcahy and LaMotta [43] observed a profound dependence of biofilm dry density on biofilm thickness in a laboratory-scale FBBR for denitrification of a municipal wastewater. Their results are shown in Fig. 6 in which the solid lines were drawn by eye to give a better illustration of the dependence of biofilm dry density on biofilm thickness. The decrease in biofilm dry density is most apparent for relatively thick biofilms. It also shows that the biofilm dry density attains a minimum value when the biofilms are overgrown. Table 2 summarizes some biofilm dry density values experimentally determined from several biofilm reactors treating different types of wastewaters.

The decrease in biofilm dry density with increasing biofilm thickness is significant in the operation of an FBBR because within the reactor stratification occurs and the less dense bioparticles with thicker biofilms tend to concentrate at the top of the reactor. This can lead to bioparticle wash-out and bed height control problems.

Table 2. Biofilm dry density values determined from different biofilm reactors

Substrate	Reactor configuration	Biofilm dry density ($g\ mL^{-1}$)	Remarks	Ref.
Glucose	Annular reactor	0.095		31)
Glucose	Tubular reactor	0.030	$S_i = 20\ mg\ L^{-1}$[a]	39)
		0.040	$S_i = 30\ mg\ L^{-1}$	
		0.050	$S_i = 70\ mg\ L^{-1}$	
Synthetic medium[b]	Tubular reactor	0.105	$\delta = 100\ \mu m$[c]	42)
		0.030	$\delta > 400\ \mu m$	
Municipal wastewater	Tubular reactor	0.050	$\delta = 100\ \mu m$	44)
		0.045	$\delta = 500\ \mu m$	
		0.030	$\delta = 1100\ \mu m$	
Municipal wastewater	Inclined plane	0.033		45)
Municipal wastewater	Trickling filter	0.077		46)
Industrial wastewater[d]	Rotating biological contractor	0.028	$\delta = 1100\text{–}3800\ \mu m$	47)
Synthetic ammonium wastewater	Biofilm[e]	0.050–0.080		48)
Nitrified secondary effluent	FBBR	0.065	$\delta \leqq 300\ \mu m$	43)
		$0.097\text{–}0.0118\delta$[f]	$300\ \mu m < \delta < 630\ \mu m$	
		0.030	$\delta \geqq 630\ \mu m$	
Corn starch wastewater	FBBR	0.072	$\delta = 150\ \mu m$	40)
		0.068	$\delta = 163\ \mu m$	
Municipal wastewater	FBBR	0.075	$\delta = 89\ \mu m$	40)
		0.048	$\delta = 113\ \mu m$	
		0.039	$\delta = 292\text{–}312\ \mu m$	

[a] S_i = inlet substrate concentration
[b] Glucose + nutrient broth + hydrated sodium acetate
[c] δ = biofilm thickness
[d] Generated from an industrial process for recovery of urban solid wastes
[e] Made by filtering microorganisms onto the support filter
[f] Equation proposed

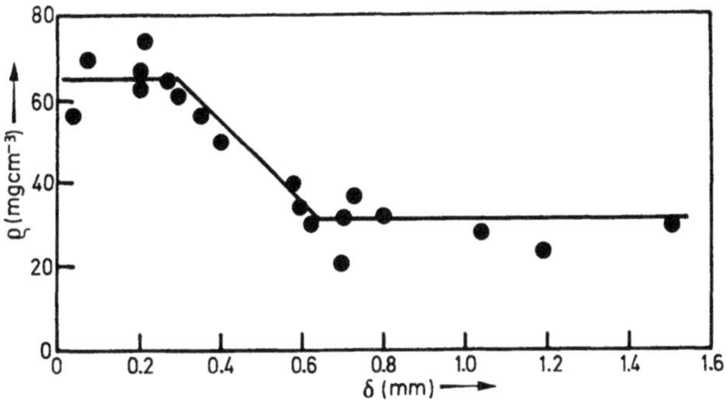

Fig. 6. Dependence of biofilm dry density (ρ) on biofilm thickness (δ) in a denitrification FBBR for a municipal wastewater [43]

Mass Transfer

The gelatinous structure of the biofilm tends to slow the transport of substrate through the biofilm and therefore, the substrate concentration surrounding the microorganisms within the biofilm is less than that in bulk liquid. Thus, the mass transport properties of the biofilm are of critical importance in assessing the overall performance of these reactors. The effective diffusivity of substrate in the biofilm is commonly used to quantify the effects of biofilm on mass transfer of substrate. Effective diffusivities for various compounds through biofilms are summarized in Table 3. Because the effective diffusivity must be measured experimentally, the reported values are expected to vary over a wide range.

Table 3. Effective diffusivities in biofilms

Substrate	Reactor configuration	Effective diffusivity (10^{-5} cm^2 s^{-1})	Ref.
Oxygen	Rotating tube	1.5	[44]
Oxygen	Plate	0.04	[49]
Oxygen	Trickling filter	0.82	[50]
Glucose	Annular reactor	0.28	[51]
Glucose	Inclined glass plate	0.07	[52]
Ammonia	Inclined glass plate	1.7	[53]
Ammonia	Biofilm[a]	1.5	[48]
Nitrate	Rotating-disk	0.82	[43]

[a] Made by filtering dispersed microorganisms onto the support filter

2.2.3 Chemical Properties of Biofilms

The inorganic composition of biofilms varies with the chemical composition of the bulk liquid and this may affect biofilm structure. Inert suspended solids may be transported to the biofilm and incorporated into its gelatinous structure in a way that may influence its chemical composition [21].

The organic composition of biofilms varies with the energy and carbon sources available in the bulk liquid for microbial metabolism and respiration. Moreover, certain elements such as nitrogen, phosphorus, and sulfur at growth-limiting concentrations in the bulk liquid stimulate the production of polysaccharide which is the building block of the biofilm.

Biofilms are highly hydrated, containing as much as 96% water. Shieh, et al.[40] have experimentally demonstrated that the moisture content of biofilms grown in FBBRs for municipal applications is approximately 0.93. They further reported that the moisture content of biofilms remains relatively constant when biofilm thickness and substrate loading rate vary over a wide range. Similar results were also reported by Mulcahy and LaMotta[43].

2.2.4 Biological Properties of Biofilms

The microorganisms which adhere to the surface strongly influence subsequent biofilm development and biofilm physical and chemical properties. Microbial populations may shift during biofilm development. In one study conducted by Alleman et al.[54] scanning electron microscopy technique was employed to identify the distribution of microbial populations in rotating biological contact (RBC) biofilms. They reported the occurrence of stratification of morphologically diverse microorganisms in the biofilms scanned, with a dense interwoven mat of filamentous microorganisms comprising the top layer and straight and comma-shaped rods in the lower strata.

LaMotta[51] measured the viability of biomass in biofilms by monitoring ATP (adenosine triphosphate) concentration with increasing biofilm thickness in a continuous-flow annular biofilm reactor inoculated with a biofilm sample from a trickling filter. Glucose was the sole energy and carbon source. As illustrated in Fig. 7, the mass of ATP increases in proportion to biofilm thickness up to a thickness of approximately 320 microns. The relatively constant ATP level at greater biofilm thickness indicates that a portion of biofilm is non-viable. The viable cell numbers, according to Characklis[39], are relatively low in relation to the total biofilm volume occupying only 1–10% of the biofilm in dilute nutrient solutions.

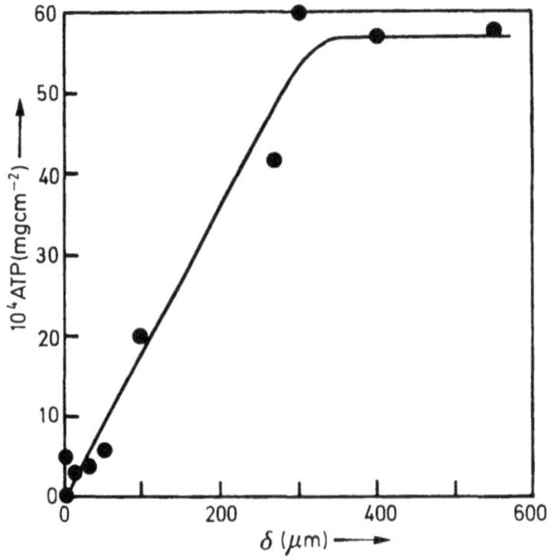

Fig. 7. Relationship between mass of ATP per unit biofilm surface area and biofilm thickness (δ) in an aerobic biofilm reactor with glucose as the sole carbon and energy source[51]

2.3 Factors Affecting Biofilm Formation

2.3.1 Polysaccharide Materials

It is believed that the attached microorganisms secrete the polysaccharide materials which hold the microorganisms together [22-24]. The polysaccharide materials secreted by microorganisms possess high absorptive properties which allow biofilm microorganisms to grow in an extremely dilute nutrient medium [55].

LaMotta et al. [20] attempted to stimulate biofilm attachment and growth on a plexiglas plate by coating the surface with synthetic and naturally produced polymers. Partial success was obtained. They concluded that surface conditioning of inert media by growing biofilms which can be easily developed is the only practical means to enhance the final biofilm growth.

2.3.2 Media Surface Characteristics

Heukelekian [56] stated that surface properties are important in initial biofilm formation, a view substantiated by the results of Jeris et al. [57] in their denitrification FBBR study. They stated that they were unable to achieve significant microbial growth on sand particles whereas good growth and excellent nitrate reduction were obtained on fresh activated carbon within two weeks of start-up.

2.3.3 Microbial Species

Heterogeneous cultures have shown an appreciable superiority over pure cultures for biofilm growth [21]. There are certain species of microorganisms which are more likely to attach and reproduce on the media surface. For instance, Characklis [21] reported that, under identical conditions, more biofilm growth was associated with the species *Bacteroides* than with the species *Bacillus*.

The biomass concentrations of FBBRs for municipal applications as shown in Table 1 provide some indication that denitrifying microorganisms are able to produce more biofilms on the fluidized media than heterogeneous microorganisms for carbon oxidation and autotrophic microorganisms for nitrification. Nevertheless, no significant differences in biofilm dry density and biofilm moisture content were observed among those FBBRs [40].

2.3.4 Hydrodynamic Environment

Greater biofilm growth is observed at higher velocities past the biofilm surface [21, 30, 56] in spite of the fact that higher fluid velocities tend to decrease the equilibrium biofilm thickness. This effect is attributed to the fact that more substrate can be transported from the bulk liquid to the biofilm surface at higher fluid velocities. Characklis [21] reported that biofilm grown under high shear stress conditions adhere to the media surface more firmly.

The hydrodynamic environment in an FBBR is somewhat different from that in conventional biofilm reactors [8, 40, 43] because of the laminar flow conditions (i.e., low Reynolds number) maintained in an FBBR. The accumulation of biofilm on the fluidized medium is more significantly affected by attrition among bioparticles than by fluid velocity past the biofilm surface. Evidence of this may be seen in Fig. 8 [8].

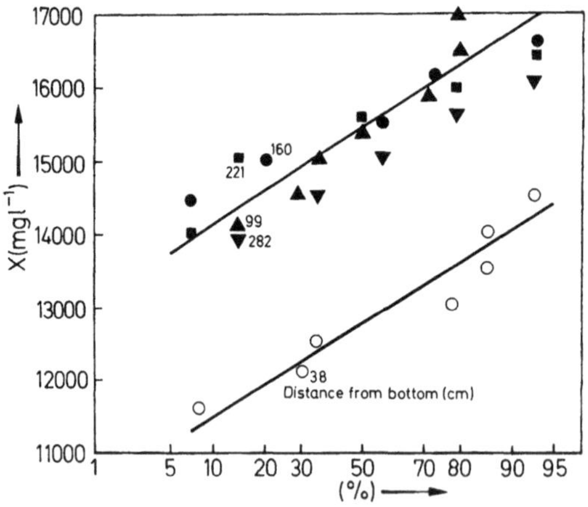

Fig. 8. Biomass concentration (X) and its distribution in an FBBR for combined carbon oxidation-nitrification of a municipal wastewater [8]. Silica sand was used as the media

The observed biomass concentrations in the lower region of the reactor are less than those observed in the upper region. This demonstrates that the attrition among bioparticles in the lower region is high enough to shear the biofilms from the media. This conclusion was verified by visual observation done under the microscope that most of the media were partially-coated. On the other hand, the biomass concentrations measured in the upper region were relatively uniform where attrition among bioparticles was minimal. This points out the significance of adequate wastewater distribution at the inlet of the reactor where much of the kinetic energy should be dissipated.

2.3.5 Substrate Characteristics and Concentration

The nature of the substrate in the bulk liquid strongly affects the production of polysaccharide in biofilms [21,33]. For instance, glucose and other sugars stimulate the growth of biofilms on the media surface because they provide "building blocks" for polysaccharide production. Atkinson and Swilley [58] reported that substrate uptake by biofilms is inhibited by the presence of inorganic colloidal material in the nutrient medium.

The availability of nutrients such as nitrogen, phosphorus, and sulfur in nutrient medium also affects the polysaccharide production [21,33]. The dry weight of biofilms has been reported to increase with increasing C/N ratio.

Heukelekian [56] observed that abundant biofilms which readily slough are typical of growth in nutrient solutions high in oxidizable materials whereas dense, readily adhering biofilms are more common in low concentration nutrient solutions.

2.3.6 Environmental Conditions

Biofilm growth is maximum when pH is maintained near neutrality. Maximum polysaccharide production by a microbial culture also occurs at neutral pH [21,33].

The reported optimal temperatures regarding biofilm attachment and growth are confusing [33]. One intersting observation indicates that polysaccharide production

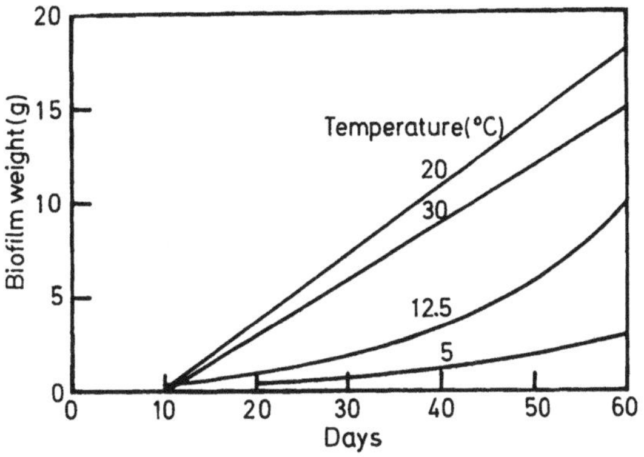

Fig. 9. Biofilm accumulation at four temperatures on vertical screens dosed with settled sewage [59]. Regression curves are illustrated

is often greater at temperatures which are sub-optimal for microbial growth. This may in part explain the reported discrepancy regarding the optimal temperature for biofilm attachment and growth.

Shieh et al. [59] reported that, in their study of biofilm growth on vertical screens dosed with settled sewage, the optimal temperature for development of biofilms lies between 20 and 30 °C (Fig. 9).

3 Process Kinetics of Fluidized Bed Biofilm Reactor

Because virtually all the biomass in an FBBR is part of a biofilm, substrates must be transported into the biofilm before they can be metabolized. The gelatinuous biofilm tends to retard substrate transport and thus causes the substrate concentration surrounding the microorganisms to be less than that in the bulk liquid. Because the rates of microbial processes are determined by the substrate concentration surrounding the microorganisms, it is essential that the combined effects of microbial rate processes and physical mass transport phenomena are considered when modelling FBBRs.

Another important consideration in modelling FBBRs is the mechanics of fluidization. For a given set of operating conditions, an analysis of the fluidization mechanics within an FBBR yields two critical parameters: the equilibrium biofilm thickness and bed porosity. This information can, in turn, be used to calculate biomass concentration. The modelling approach commonly employed for heterogeneous catalytic processes is employed here to derive an overall FBBR kinetic model.

3.1 Mass Transfer Limitations — Concept of Effectiveness Factor

Substrate conversion in a heterogeneous bioreactor such as an FBBR can be described by the following steps [51]:

1) Transport of substrate from the bulk liquid to the liquid-biofilm interface (external mass transfer);

2) Transport of substrate within the biofilm (internal mass transfer), and
3) Substrate conversion reactions within the biofilm.

Steps 2 and 3 take place simultaneously and thus neither can be said to control while step 1 occurs in series with steps 2 and 3. For intrinsic reaction rates with positive dependence on substrate concentration (i.e., first order, Michaelis-Menten, etc.) the gradients established by steps 1 and 2 decrease the observed reaction rate by decreasing local (i.e., intrabiofilm) substrate concentration. For intrinsic zero order kinetics, steps 1 and 2 can decrease the observed reaction rate by limiting substrate penetration into the biofilm.

3.1.1 External Mass Transfer

The significance of external mass transfer in the FBBR can be assessed using a correlation developed for fluidized beds [43]:

$$k_c = \frac{0.81}{\varepsilon}\left[\frac{D^{4/3}\,U\varrho_l^{1/3}}{\mu^{1/3}\,d_p}\right]^{1/2} \tag{1}$$

where k_c is the external mass transfer coefficient, LT^{-1}; ε is the bed porosity; D is the diffusivity of substrate in the bulk liquid, L^2T^{-1}; U is the superficial upflow velocity, LT^{-1}; ϱ_l is the liquid density, ML^{-3}; μ is the dynamic viscosity of the liquid, MTL^{-2}; and d_p is the bioparticle diameter, L.

Eq. (1) was selected because of its derivation from experimentation at Reynolds numbers within the range common to FBBR operation. A typical FBBR external mass transfer coefficient calculated by Eq. (1) is $k_c = 0.01$ cm s^{-1}. Using this coefficient and intrinsic biofilm denitrification parameters determined by Mulcahy, et al. [37], the effect of external mass transfer on the observed FBBR denitrification rate was calculated for a range of bulk-liquid substrate (nitrate) concentrations. The results are summarized in Table 4. As can be seen from Table 4, errors in the observed rate which result if external mass transfer effects are neglected, range from 2.6 to 7.1% for bulk-liquid NO_3^-–N concentrations over the range 43 to 6 mg L^{-1}. Errors of this magnitude are acceptable in light of the greatly simplified mathematics which result. Thus, external mass transfer will be neglected here in the development of the FBBR kinetic model.

Table 4. Effect external mass transfer on bioparticle observed reaction rate [37]

Biofilm surface		Bulk liquid		% Error in rate
Concentration (mg L^{-1})	Rate × 10^{10} (g s^{-1})	Concentration (mg L^{-1})	Rate × 10^{10} (g s^{-1})	
40	8.7	42.4	9.0	2.6
30	7.7	32.0	7.9	2.9
20	6.4	21.7	6.6	3.6
10	4.7	11.2	4.9	5.0
5	3.4	5.9	3.7	7.1

Intrinsic Denitrification Parameters
$k_0 = 3{,}32 \times 10^{-5}\ s^{-1}$
$D = 0.815 \times 10^{-5}\ cm^2\ s^{-1}$
$\varrho = 62.1$ mg cm^{-3}

Reactor Conditions
$U = 1.07$ cm s^{-1}
$\delta = 0.0207$ cm
$d_m = 0.0682$ cm

3.1.2 Effectiveness Factor Expressions

The effect of mass transfer on the intrinsic reaction rate is commonly quantified by an effectiveness factor [51]. Under the assumption of negligible external mass transfer, the effectiveness factor (η) is defined as:

$$\eta = \frac{\text{observed reaction rate}}{\text{intrinsic reaction rate at bulk-liquid conditions}} \tag{2}$$

Mathematical expressions for a bioparticle effectiveness factor are developed subject to the following constraints:
1) Homogeneous biofilm of uniform thickness;
2) Spherical media of uniform size;
3) Negligible external mass transfer resistance;
4) Internal mass transfer described by Fick's first law;
5) Single limiting soluble substrate; and
6) Steady state conditions.

The bioparticle continuity equation for the limiting substrate is then:

$$\frac{D_e}{r^2}\frac{d}{dr}\left(r^2\frac{dS}{dr}\right) = R \tag{3}$$

where D_e is the effective diffusivity of substrate in the biofilm, L^2T^{-1}; r is the radial distance measured from the bioparticle center, L; S is the intrabiofilm substrate concentration, ML^{-3}; and R is the observed substrate conversion rate per unit biofilm mass, $MM^{-1}T^{-1}$.

Zero Order Kinetics

For intrinsic zero order kinetics, the following reaction rate and boundary conditions apply (Fig. 10a):

$$R = \varrho k_0 \tag{4}$$

$$S = S_b \qquad \text{at} \qquad r = r_p \tag{5}$$

$$\frac{dS}{dr} = S = 0 \qquad \text{at} \qquad r = r_c \tag{6}$$

where ϱ is the biofilm dry density, ML^{-3}; k_0 is the intrinsic zero order rate constant, $MM^{-1}T^{-1}$; S_b is the bulk-liquid substrate concentration, ML^{-3}; r_p is the bioparticle radius, L; and r_c is the substrate penetration depth, L.

Integration of Eqs. (3) and (4) subject to the above boundary conditions yields:

$$\left(\frac{r_c}{r_p}\right)^3 - 1.5\left(\frac{r_c}{r_p}\right)^2 + \left(\frac{1}{2} - \frac{3}{\Phi_0^2}\right) = 0 \tag{7}$$

where $\Phi_0 = r_p\left(\frac{\varrho k_0}{D_e S_b}\right)^{1/2}$, the conventional zero order Thiele modulus.

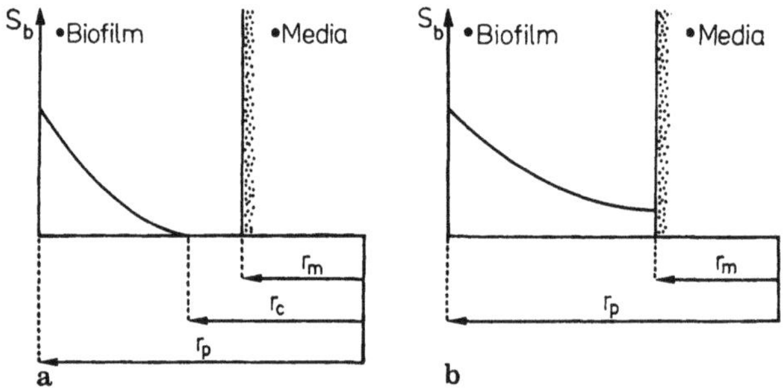

Fig. 10 a and b. (a) Boundary conditions for intrinsic zero order kinetics; and (b) boundary conditions for intrinsic first order kinetics.
S_b: bulk-liquid substrate concentration
r_m: media radius
r_c: substrate penetration depth
r_p: bioparticle radius

The effectiveness factor for the intrinsic zero order reaction (η_0) is the ratio of biofilm volume with substrate concentration greater than zero to the total biofilm volume [59]:

$$\eta_0 = \frac{1 - (r_c/r_p)^3}{1 - (r_m/r_p)^3} \tag{8}$$

The functional relationship among η_0, Φ_0 and (r_m/r_p) is illustrated in Fig. 11 [59].

The functional relationship illustrated in Figure 11 can be simplified by replacing the bioparticle radius of the conventional zero order Thiele modulus with a characteristic radius ($\bar{r}$), defined as:

$$\bar{r} = \frac{\text{biofilm volume}}{\text{biofilm exterior surface area}} \tag{9}$$

$$\bar{r} = \frac{(r_p^3 - r_m^3)}{3r_p^2} \tag{10}$$

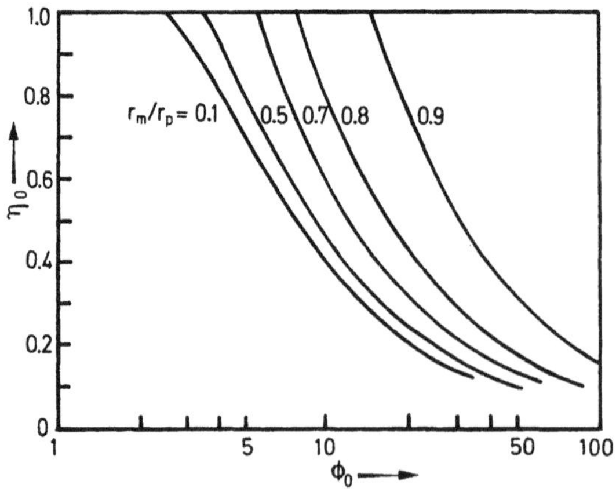

Fig. 11. Bioparticle zero order effectiveness factor (η_0) versus conventional zero order Thiele modulus (Φ_0) for various ratios of media radius (r_m) to bioparticle radius (r_p)

A modified zero order Thiele modulus (Φ_{0m}) is then defined as:

$$\Phi_{0m} = \bar{r}\left(\frac{\varrho k_0}{D_e S_b}\right)^{0.5} \tag{11}$$

Numerical solution of Eqs. (8), (10) and (11) yields a linear log-log relationship as shown in Fig. 12 which is well described by the following empirical Eq. [59]:

$$\eta_0 = 1.2712\Phi_{0m}^{-1.0} = \frac{3.814}{\left[1 - \left(\frac{r_m}{r_p}\right)^3\right]}\left(\frac{D_e}{\varrho k_0 r_p^2}\right)^{0.5} S_b^{0.5} \tag{12}$$

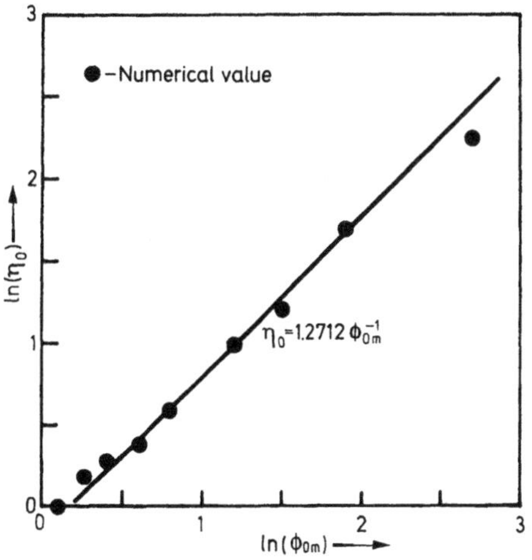

Fig. 12. Bioparticle zero order effectiveness factor (η_0) versus modified zero order Thiele modulus (Φ_{0m})

An explicit relationship is thus established for the intrinsic zero order reaction in partially penetrated bioparticles which shows the effectiveness factor to be proportional to the bulk-liquid substrate concentration to the 0.50 power. Setting $\eta_0 = 1.0$ in Eq. (12) gives $\Phi_{0m} = 1.27$, the value at which transition occurs from full to partial substrate penetration of the biofilm. For $\Phi_{0m} \leqq 1.27$, the whole biofilm is active and intrinsic zero order kinetics are observed. For $\Phi_{0m} > 1.27$, the inner portion of the biofilm is inactive and the observed reaction rate is proportional to $S_b^{0.50}$.

First Order Kinetics

The intrinsic first order reaction rate and boundary conditions for Eq. (3) are as follows (Fig. 10b):

$$R = \varrho k_1 S \tag{13}$$

$$S = S_b \qquad \text{at } r = r_p \tag{14}$$

$$\frac{dS}{dr} = 0 \qquad \text{at } r = r_m \tag{15}$$

where k_1 is the intrinsic first order rate constant, $L^3M^{-1}T^{-1}$; and r_m is the media radius, L.

Integration of Eqs. (3) and (13) subject to the above boundary conditions yields:

$$S = \frac{S_b r_p}{r}\left[\frac{(\alpha r_m + 1)\, e^{\alpha(r - r_m)} + (\alpha r_m - 1)\, e^{-\alpha(r - r_m)}}{(\alpha r_m + 1)\, e^{\alpha\delta} + (\alpha r_m - 1)\, e^{-\alpha\delta}}\right] \tag{16}$$

where $\alpha = (\varrho k_1/D_e)^{0.5}$; and $\delta = r_p - r_m$, the biofilm thickness, L.

Under steady state conditions, the observed substrate conversion rate by a bioparticle is equal to the mass transfer rate of substrate across the liquid-biofilm interface, $W|_{r=r_p}$, expressed as follows:

$$W|_{r=r_p} = -4\pi r_p^2 D_e \left.\frac{dS}{dr}\right|_{r=r_p}$$

$$= -4\pi S_b r_p D_e \left\{\frac{(\Phi_1 - 1)\left[1 + \Phi_1\left(\frac{r_m}{r_p}\right)\right] e^{\alpha\delta} + (\Phi_1 + 1)\left[1 - \Phi_1\left(\frac{r_m}{r_p}\right)\right] e^{-\alpha\delta}}{\left[\Phi_1\left(\frac{r_m}{r_p}\right) + 1\right] e^{\alpha\delta} + \left[\Phi_1\left(\frac{r_m}{r_p}\right) - 1\right] e^{-\alpha\delta}}\right\} \tag{17}$$

where $\Phi_1 = \alpha r_p = r_p(\varrho k_1/D_e)^{0.5}$, the conventional first order Thiele modulus.

The bioparticle first order effectiveness factor (η_1) is then:

$$\eta_1 = \frac{W|_{r=r_p}}{-\frac{4}{3}\pi\,(r_p^3 - r_m^3)\,\varrho k_1 S_b}$$

$$= \frac{\Phi_1^2}{3\left[1 - \left(\frac{r_m}{r_p}\right)^3\right]}\left\{\frac{(\Phi_1 - 1)\left[1 + \Phi_1\left(\frac{r_m}{r_p}\right)\right] e^{\alpha\delta} + (\Phi_1 + 1)\left[1 - \Phi_1\left(\frac{r_m}{r_p}\right)\right] e^{-\alpha\delta}}{\left[\Phi_1\left(\frac{r_m}{r_p}\right) + 1\right] e^{\alpha\delta} + \left[\Phi_1\left(\frac{r_m}{r_p}\right) - 1\right] e^{-\alpha\delta}}\right\} \tag{18}$$

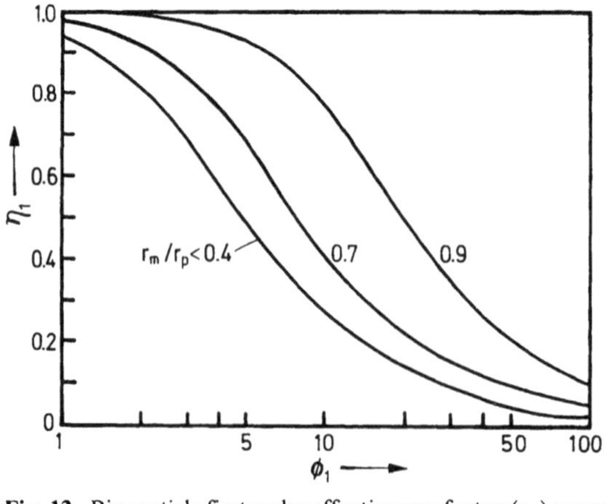

Fig. 13. Bioparticle first order effectiveness factor (η_1) versus conventional first order Thiele modulus (Φ_1) for various ratios of media radius (r_m) to bioparticle radius (r_p)

The functional relationship among η_1, Φ_1, and (r_m/r_p) is presented in Fig. 13 [59].

Following the approach used previously to obtain a single, explicit expression for the effectiveness factor, a first order modified Thiele modulus (Φ_{1m}) is defined in terms of the characteristic radius ($\bar{r}$) as follows:

$$\Phi_{1m} = \left(\frac{\bar{r}}{r_p}\right) \Phi_1 \tag{19}$$

Numerical solution of Eqs. (18) and (19) for different (r_m/r_p) values gives the data points shown in Fig. 14 which is well described by the first order, non-spherical form advanced by Aris [60] for homogeneous reaction media:

$$\eta_1 = \frac{\coth(3\Phi_{1m})}{\Phi_{1m}} - \frac{1}{3\Phi_{1m}^2} \tag{20}$$

From Eqs. (19) and (20), it is clear that the first order effectiveness factor is a function of bioparticle size only and is independent of bulk-liquid substrate concentration.

Setting $\eta_1 = 1.0$ in Eq. (20) gives $\Phi_{1m} = 0.11$, the value at which transition occurs from observable intrinsic kinetics to a regime in which the observed reaction rate is limited by internal mass transfer.

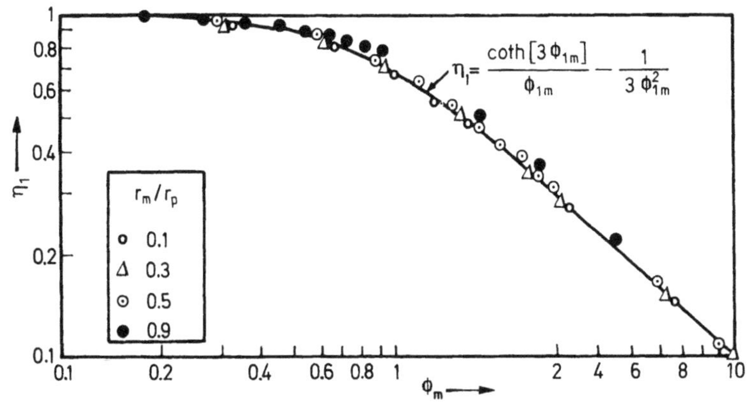

Fig. 14. Bioparticle first order effectiveness factor (η_1) versus modified first order Thiele modulus (Φ_{1m})

3.2 Fluidization Mechanics

The main difference between conventional biofilm reactors and FBBRs is the free movement of bioparticles in the latter. Moreover, the growth of biofilm changes the overall density of the bioparticles and therefore, the expansion of the fluidized bed. Thus, the number and size of bioparticles (and therefore, biofilm thickness) per unit fluidized bed volume provide two critical pieces of information for the estimation of FBBR biomass concentration. A logical and convenient way to obtain this information is to define the fluidization mechanics in terms of measurable parameters. The resulting correlations would provide a basis for calculating equilibrium biofilm thickness (δ), bed porosity (ε) or degree of bed expansion and biomass concentration (X) which correspond to a given set of operating conditions [12,61,62].

Several solid-liquid fluidization correlations, which link particle concentration in the fluidized state to the physical characteristics of a fluidized bed reactor, have been developed in past decades [63-66]. Among them the Richardson-Zaki correlation is widely used:

$$\frac{U}{U_t} = \varepsilon^n \tag{21}$$

where U_t is the bioparticle terminal setting velocity, LT^{-1}; and n is the expansion index.

For a smooth, spherical bioparticle (Fig. 1), the bioparticle terminal settling velocity may be calculated using Newton's law [43]:

$$U_t = \left[\frac{0.75(\varrho_s - \varrho_l)\, g d_p}{C_D \varrho_l}\right]^{1/2} \tag{22}$$

where ϱ_s is the bioparticle density, ML^{-3}; g is the gravitational acceleration, LT^{-2}; and C_D is Newton's drag coefficient.

Correlations for drag coefficient (C_D) and expansion index (n) have been developed empirically for the FBBR [43]:

$$C_D = 36.67\, N_{Re_t}^{-0.67}, \qquad 40 < N_{Re} < 90 \tag{23}$$

$$n = 10.35\, N_{Re_t}^{-0.18}, \qquad 40 < N_{Re} < 90 \tag{24}$$

where $N_{Re_t} = (d_p \varrho_l U_t/\mu)$ is the terminal Reynolds number.

The bioparticle density (ϱ_s) is:

$$\varrho_s = \varrho_m \left(\frac{r_m}{r_p}\right)^3 + \frac{\varrho}{(1 - P)}\left[1 - \left(\frac{r_m}{r_p}\right)^3\right] \tag{25}$$

where ϱ_m is the media density, ML^{-3}; and P is the biofilm moisture content.

Using Eqs. (21) to (25), the biomass concentration in an FBBR is calculated as:

$$X = \varrho(1 - \varepsilon)\left[1 - \left(\frac{r_m}{r_p}\right)^3\right] = \frac{\varrho V_m}{AH_B}\left[\left(\frac{r_p}{r_m}\right)^3 - 1\right] \tag{26}$$

where X is the biomass concentration in an FBBR, ML^{-3}; V_m is the media volume, L^3; A is the cross-sectional area of the reactor, L^2; and H_B is the expanded bed height, L.

Prediction of biomass concentration in an FBBR utilizing the correlations described herein can be facilitated via the iterative procedure as schematically illustrated in Fig. 15. For a specific application, an estimated value of biofilm thickness, $\hat{\delta}$, and relevant input information (i.e., U, A, V_m, etc.) are substituted into the fluidization

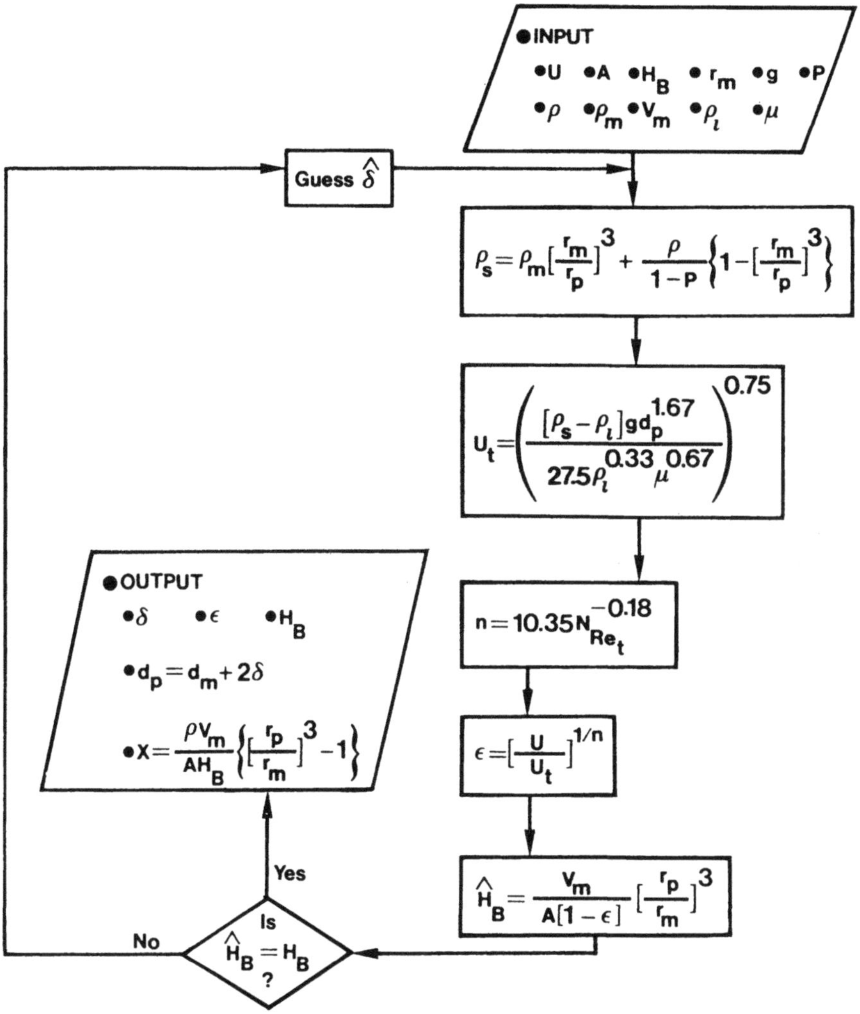

Fig. 15. Fluidization algorithm

correlations stepwise as shown in Fig. 15. From these correlations, an estimated expanded bed height, $\hat{H}_B$, is calculated. If $\hat{H}_B$ equals the H_B value, the estimated biofilm thickness $\hat{\delta}$, is the desired design and operating value and the corresponding biomass concentration can be calculated. Otherwise, a new $\hat{\delta}$ is chosen to calculate another H_B value.

3.3 Overall Rates of Substrate Conversion

Because recycle of reactor effluent is sometimes employed in the operation of an FBBR to insure uniform fluidization and an adequate substrate loading rate, the overall rate expressions for substrate conversion are strongly affected by FBBR hydraulic characteristics. It has been experimentally demonstrated, via tracer studies, that an

FBBR can be treated as a plug-flow reactor as long as the recycle ratio is kept below 2. Otherwise, a completely-mixed flow model must be used [7].

3.3.1 Plug-Flow Conditions (Recycle Ratio $\leqq 2$)

A mass balance on the bulk-liquid substrate concentration across the FBBR illustrated in Fig. 1 yields:

$$U \frac{dS_b}{dZ} + R_v = 0 \tag{27}$$

with boundary conditions:

$$S_b = S_{bi} \qquad \text{at} \qquad Z = 0 \tag{28}$$

where R_v is the observed substrate conversion rate per unit fluidized bed volume, $ML^{-3}T^{-1}$; S_{bi} is the inlet substrate concentration at $Z = 0$, ML^{-3}; and Z is the axial position, L.

The reaction term (R_v) in Eq. (27) is more usefully expressed as:

$$R_v = RX \tag{29}$$

Zero Order Kinetics

For intrinsic zero order kinetics:

$$R_v = RX = \eta_0 k_0 X = \frac{3.814}{\left[1 - \left(\frac{r_m}{r_p}\right)^3\right]} \left(\frac{k_0 D_e}{\varrho r_p^2}\right)^{0.5} X S_b^{0.5} \tag{30}$$

Integration of Eq. (27) subject to Eqs. (28) and (30) yields:

$$S_b^{0.5} = S_{bi}^{0.5} - K_0 Z \tag{31}$$

where $K_0 = \frac{1.907}{\left[1 - \left(\frac{r_m}{r_p}\right)^3\right]} \left(\frac{k_0 D_e}{\varrho r_p^2}\right)^{0.5} \left(\frac{X}{U}\right)$, an observed zero order rate constant, $M^{0.5}L^{-2.5}$

Thus, the substrate concentration profile through an intrinsic zero order FBBR limited by internal mass transfer is described by a half order equation. The value of K_0 will vary, depending on the characteristics of both substrate and biofilm and on the prevailing operating conditions.

First Order Kinetics

For intrinsic first order kinetics:

$$R_v = \eta_1 k_1 S_b X = \left\{ \frac{\coth\left[3\bar{r}\left(\frac{\varrho k_1}{D_e}\right)^{0.5}\right]}{\bar{r}\left(\frac{\varrho k_1}{D_e}\right)^{0.5}} - \frac{1}{\left(\frac{3\bar{r}\varrho k_1}{D_e}\right)} \right\} k_1 S_b X \tag{32}$$

Integration of Eq. (27) subject to Eqs. (28) and (32) yields:

$$S_b = S_{bi}\, e^{-K_1 Z} \tag{33}$$

where $K_1 = \frac{X}{\varrho \bar{r} U}\left\{(\varrho k_1 D_e)^{0.5} \coth\left[3\bar{r}\left(\frac{\varrho k_1}{D_e}\right)^{0.5}\right] - \frac{D_e}{3}\right\}$, an observed first order rate constant, L^{-1}.

Thus, the substrate concentration profile through an intrinsic first order FBBR limited by internal mass transfer is described by an exponential equation. The value of K_1 will vary as a function of substrate and biofilm as well as existing operating conditions.

3.3.2 Completely Mixed Conditions (Recycle Ratio >2)

A mass balance on the bulk-liquid substrate concentration across the FBBR illustrated in Fig. 1 yields:

$$\frac{(S_{bi} - S_b)}{\theta} = RX \tag{34}$$

where $\theta = H_B/U$, the hydraulic retention time, T.

Zero Order Kinetics

For intrinsic zero order kinetics:

$$\frac{(S_{bi} - S_b)}{\theta} = \frac{3.814}{\left[1 - \left(\frac{r_m}{r_p}\right)^3\right]}\left(\frac{k_0 D_e}{\varrho r_p^2}\right)^{0.5} X S_b^{0.5} \tag{35}$$

or

$$S_b = S_{bi} + 2(K_0 H_B)^2 - 2(K_0 H_B)\sqrt{S_{bi}} + (K_0 H_B)^2 \tag{36}$$

First Order Kinetics

For intrinsic first order kinetics:

$$\frac{(S_{bi} - S_b)}{0} = \left\{\frac{\coth\left[3\bar{r}\left(\frac{\varrho k_1}{D_e}\right)^{0.5}\right]}{\bar{r}\left(\frac{\varrho k_1}{D_e}\right)^{0.5}} - \frac{1}{\left(\frac{3\bar{r}\varrho k_1}{D_e}\right)}\right\} k_1 S_b X \tag{37}$$

or

$$S_b = \frac{S_{bi}}{(1 + K_1 H_B)} \tag{38}$$

3.4 Limiting Bulk-Liquid Substrate Concentration

The effectiveness factor expressions defined above provide a convenient way of examining the effects of bulk-liquid substrate concentration on the observed substrate conversion rate in an FBBR. Effectiveness factor expressions shown in Eqs. (12)

and (20) indicate that the observed substrate conversion rate is only affected by the bulk-liquid substrate concentration in an intrinsic zero order FBBR. Such effects are shown in Fig. 16 for a denitrification FBBR, an intrinsic zero order system [37,67]. An $\eta_0 = 0.90$ as recommended by Kobayashi and Laidler [68] is employed in the preparation of Fig. 16 [37].

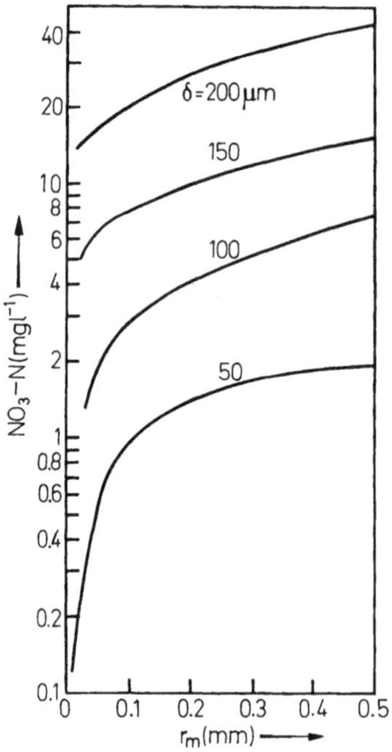

Fig. 16. Limiting bulk-liquid nitrate-nitrogen concentration for full biofilm penetration (i.e., $\eta_0 = 0.90$).
k_0: $3.32 \times 10^{-5}\ s^{-1}$
D_e: $0.815 \times 10^{-5}\ cm^2\ s^{-1}$
ϱ: $65\ mg\ cm^{-3}$
P: 0.93

It is seen from Fig. 16 that the limiting bulk-liquid substrate concentration which induces significant internal mass transfer effects in an intrinsic zero order FBBR (i.e., $\eta_0 = 0.90$) is directly proportional to biofilm thickness. Higher bulk-liquid substrate concentration is required to penetrate into thick biofilm to insure that a majority of the biofilm is not substrate-starved. For instance, complete penetration of a 200 micron biofilm grown on 0.4 mm diameter media will occur when the bulk-liquid nitrate-N concentration is equal to or greater than 23 mg L^{-1}. For nitrate-N values less than 23 mg L^{-1}, only partial nitrate penetration will occur. From Fig. 16 it is evident that the observed denitrification rate in an FBBR is likely to be mass transfer limited for nitrate levels commonly found in nitrified municipal wastewaters.

3.5 Critical Biofilm Thickness

A parameter which is indirectly controllable in the operation of an FBBR is the equilibrium biofilm thickness. For a given medium, equilibrium biofilm thickness is dependent on superficial upflow velocity, expanded bed height, and media volume. Biofilm thickness is, in turn, the single most important parameter affecting biomass effectiveness and thus the overall performance of an FBBR [43].

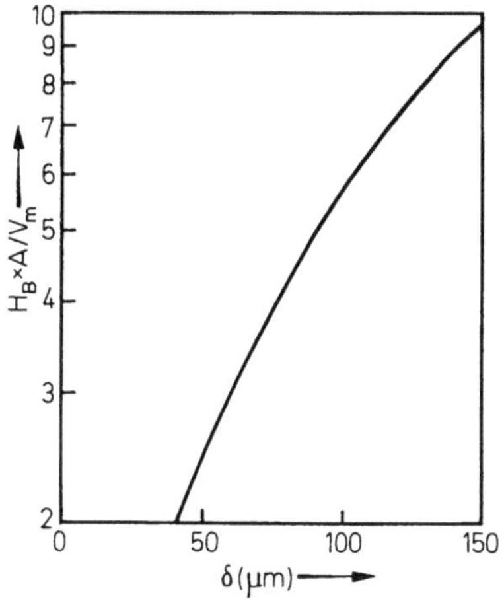

Fig. 17. Effect of biofilm thickness (δ) on FBBR bed expansion.
r_m: 0.2 mm
ϱ_m: 2.65 g cm^{-3}
U: 1 cm s^{-1}
ϱ: 65 mg cm^{-3}
P: 0.93

The effect of biofilm thickness on the bed expansion of an FBBR under a given set of operating conditions is illustrated in Fig. 17. Excessive accumulation of biofilm on the fluidized medium leads to a greater bed height increase which reduces the bed stability against hydraulic load variations. In extreme cases, wash-out of bioparticles from the reactor could occur.

The fluidization algorithm illustrated in Fig. 15 can be used to examine the effect of biofilm thickness on biomass concentration in an FBBR, as shown in Fig. 18. The parabolic shape of this curve indicates that, to the left of the maximum (i.e., 100 micron biofilm), the rate of biomass increase due to biofilm accumulation is faster than the rate of biomass decrease due to the reduction in bioparticles per unit fluidized volume caused by bed expansion. To the right; the opposite holds. If the objective is simply to maximize total biomass concentration, then the FBBR should

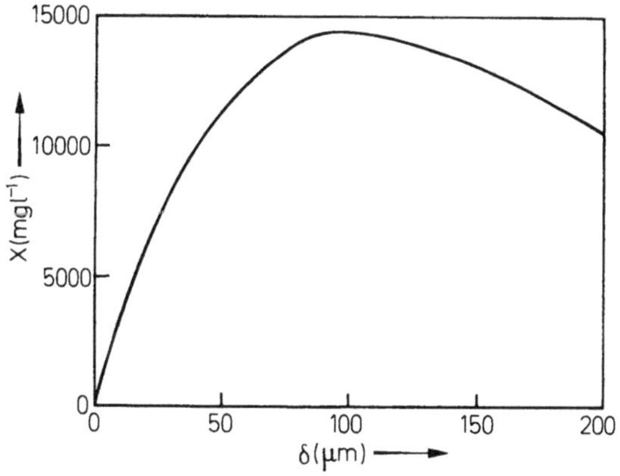

Fig. 18. Effect of biofilm thickness (δ) on FBBR biomass concentration (X).
r_m: 0.2 mm
ϱ_m: 2.65 g cm^{-3}
U: 1 cm s^{-1}
ϱ: 65 mg cm^{-3}
P: 0.93

be operated to control biofilm thickness at approximately 100 microns. Because of internal mass transfer limitations, however, this simplistic approach may not optimize overall FBBR performance. To examine this possibility, it is necessary to determine the effect of biofilm thickness on the biofilm effectiveness factor.

A concept which aids in understanding FBBR performance characteristics is that of effective biomass concentration (X_a). The effective biomass concentrations is less than X because of internal mass transfer limitations:

$$X_a = \eta X \tag{39}$$

The effective biomass concentration is, in turn, related to the overall reaction term R_v as follows:

$$R_v = k_0 X_a \quad \text{(zero order kinetics)} \tag{40}$$

or

$$R_v = k_1 X_a S_b \quad \text{(first order kinetics)} \tag{41}$$

Therefore, the observed substrate conversion rate at any point in an FBBR is directly proportional to the effective biomass concentration at the point. Maximization of effective biomass concentration will therefore maximize the observed substrate conversion rate in the FBBR. The evidence of this may be seen in Fig. 19 in which the

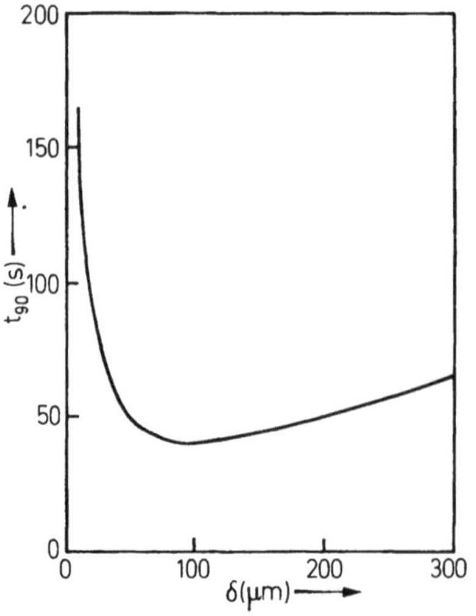

Fig. 19. Effect of biofilm thickness (δ) on the observed FBBR substrate (NO_3-N) conversion rate. t_{90} is the required reaction time for 90% NO_3-N removal.

r_m 0.2 mm
ϱ_m: 2.65 g cm^{-3}
U: 1 cm s^{-1}
ϱ: 65 mg cm^{-3}
P: 0.93
k_0: 3.32×10^{-5} s^{-1}
D_e: 0.815×10^{-5} cm^2 s^{-1}
S_{bi}: 20 mg NO_3-N L^{-1}

effect of biofilm thickness on the required reaction time for 90% NO_3^-—N removal in a plug-flow FBBR in plotted. This figure clearly shows that optimal FBBR substrate conversion does not occur at maximum total biomass concentration (i.e. 100 micron biofilm) or at maximum effectiveness factor (i.e. 50 micron biofilm), but rather at maximum effective biomass concentration throughout the reactor (i.e. 80 micron biofilm).

3.6 Critical Media Size and Density

The effect of medium characteristics on FBBR bed expansion is shown in Figs. 20 and 21. These figures show that FBBR bed height stability to variations in superficial upflow velocity increases with increasing medium size and density. This is extremely

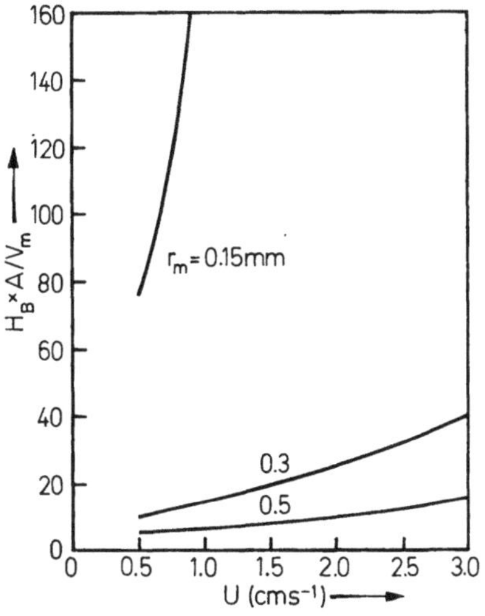

Fig. 20. Effect of media size (in terms of media radius, r_m) on FBBR bed expansion.
ϱ_m: 2.65 g cm^{-3}
δ: 100 μm
ϱ: 65 mg cm^{-3}
P: 0.93

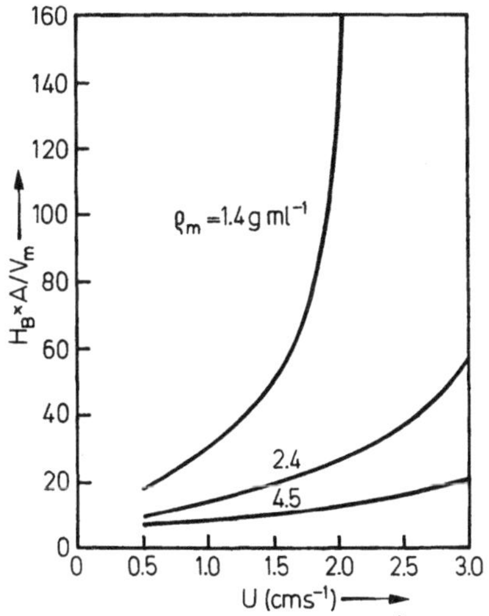

Fig. 21. Effect of media density (ϱ_m) on FBBR bed expansion.
r_m: 0.035 cm
δ: 100 μm
ϱ: 65 mg cm^{-3}
P: 0.93

important with regard to flow equalization requirements for a system utilizing an FBBR. Nevertheless, any such advantages must be weighed against the higher energy requirements for fluidization of dense and/or large media.

Similar conclusions can be drawn on the effect of medium size on the FBBR biomass concentration as illustrated in Fig. 22. The importance of maximization of effective biomass concentration rather than total biomass concentration is clearly seen in Fig. 23 in that the medium size which gives maximum biomass concentration is not necessarily the same one which gives optimal FBBR performance.

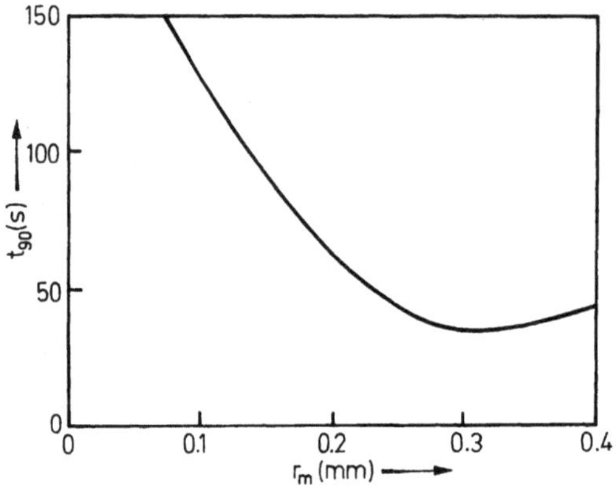

Fig. 22. Effect of media size (in terms of media radius, r_m) on the observed FBBR substrate (NO_3-N) conversion rate. t_{90} is the required reaction time for 90% NO_3-N removal.

ϱ_m: 2.65 g cm^{-3}	δ: 100 μm
U: 1 cm s^{-1}	k_0: 3.32×10^{-5} s^{-1}
ϱ: 65 mg cm^{-3}	D_e: 0.815×10^{-5} cm^2 s^{-1}
P: 0.93	S_{bi}: 20 mg NO_3-N L^{-1}

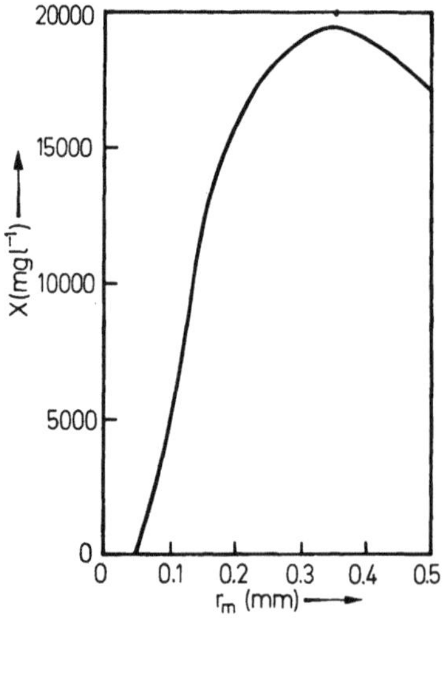

Fig. 23. Effect of media size (in terms of media radius, r_m) on FBBR biomass concentration (X).
ϱ_m: 2.65 g cm^{-3}
U: 1 cm s^{-1}
ϱ: 65 mg cm^{-3}
P: 0.93
δ: 100 μm

3.7 Parameter Estimation

3.7.1 Zero Order Kinetics

The intrinsic constants k_0 and D_e can be determined experimentally by a series of runs in a single FBBR, or preferably, by parallel runs in a series of FBBRs. In either case, the quantity to be varied from run ro run is the equilibrium biofilm thickness. This is accomplished by varying media volume in FBBRs while maintaining constant superficial upflow velocities and expanded bed heights. The objective is to achieve biofilms thin enough to allow the intrinsic rate constant (k_0) to be measured directly. Intrinsic kinetics (negligible mass transfer limitations) are observed when k_0, calculated via Eq. (31) or (35), does not increase with increasing biofilm thickness.

The effective diffusivity (D_e) can then be determined for the rest of the experimental runs (i.e., those in which observed k_0 decreases with increasing biofilm thickness) by applying Eq. (31) or (35) with K_0 and k_0 known. Because the effective diffusivity is a constant, averaging can be used to obtain a representative D_e value [69].

3.7.2 First Order Kinetics

As with the zero order case, a series of FBBR experimental runs is performed at various biofilm thicknesses. Intrinsic kinetics is observed when no increase in the quantities, k_1 and η_1, is observed with decreased biofilm thickness. Under these conditions η_1 is unity and k_1 can be calculated directly via Eq. (33) or (37). The remaining runs with greater biofilm thicknesses are used to determine effective diffusivity. From the slope of either the ln (S_b) versus ln (Z) plot or the $(S_{bi} - S_b)/\theta$ versus S_b plot, $\eta_1 k_1$ values are calculated. The corresponding η_1 values are obtained by factoring out k_1. The effective diffusivity is obtained stepwise by first calculating Φ_{1m} at a given η_1 via trial and error solution of Eq. (20). Then D_e is calculated at given Φ_{1m} and k_1 via Eq. (19).

3.7.3 Fluidization Correlations

To predict the FBBR biofilm thickness and biomass concentration under conditions which are outside the Reynolds number range covered by Eqs. (23) and (24), additional fluidization experimentation to extend and refine these correlations is recommended.

A series of runs is required to cover a broad range of biofilm thickness (and hence Reynolds number) for each medium type under consideration. For each run, superficial upflow velocity is varied (start at highest value and decrease stepwise) and the resultant bed porosity measured. Bed porosity is linked to the readily measurable bed height as follows: At the start of a run, carefully remove a volume of bioparticles from the FBBR; note the change in bed height (ΔH_B); determine the bioparticle radius; remove the biofilm from the media; clean, dry and weigh the media; and, determine the media volume in $\Delta H_B (\Delta V_m)$ by dividing media weight by media density. Total bioparticle volume in $\Delta H_B (\Delta V_s)$ is then calculated as:

$$\Delta V_s = \Delta V_m \left(\frac{r_p}{r_m}\right)^3 \tag{42}$$

Total bioparticle volume remaining in the FBBR (V_s) is:

$$V_s = \Delta V_s \left(\frac{H_B}{\Delta H_B}\right) \tag{43}$$

The bed porosity-bed height relationship is then:

$$\varepsilon = 1 - \frac{V_s}{H_B A} \tag{44}$$

Plotting ln (U) versus ln (ε) according to Eq. (21) for each run, the expansion index is obtained as the slope, and the bioparticle terminal settling velocity is obtained as the intercept at $\varepsilon = 1.0$. The bioparticle terminal settling velocity is, in turn, related to bioparticle physical parameters and drag coefficient as shown by Eq. (22).

FBBR correlations similar to Eqs. (23) and (24) can, therefore, be developed. With these correlations, the bed expansion corresponding to a specific set of bioparticle characteristics and a given superficial upflow velocity can be determined. Iterative solution then yields the equilibrium biofilm thickness and biomass concentration as illustrated in Fig. 15.

4 Design Considerations of Fluidized Bed Biofilm Reactor

4.1 Distribution of Incoming Wastewater

Two important design aspects regarding distribution of the wastewater flow coming into an FBBR should be closely examined. First, uniform distribution of incoming wastewater over the cross-sectional area of reactor perpendicular to the flow should always be maintained to avoid sprouting of the fluidized bed. Sprouting of the fluidized bed tends to disrupt the distribution pattern of bioparticles and thus substantially increase the collision frequency among bioparticles. Moreover, sprouting of the fluidized bed also intensifies the degree of short-circuiting through the reactor. In extreme cases, a portion of the reactor bottom will be packed with non-fluidized media. Second, the kinetic energy associated with incoming wastewater should be largely dissipated at the reactor bottom to insure that turbulence in the region is minimal. Otherwise, full growth of biofilms on the fluidized media will be impossible.

For small FBBRs, the wastewater is introduced upwards into the reactor. A layer of gravel or other coarse materials is placed at the reactor bottom as a buffer zone to dissipate the wastewater kinetic energy. Flow distribution devices such as perforated plates are placed above the gravel level to insure uniform distribution of wastewater over the cross-sectional area perpendicular to flow.

A system of pipe laterals is used in large FBBRs to introduce wastewater downwards into a flow distribution device located at the reactor bottom, which consists of a number of inverted conical-shaped substructures. Wastewater delivered into each inverted conical-shaped substructure makes a 180° turn with much of its kinetic energy dissipated. The layout of pipe lateral systems and the number of inverted conical-shaped substructures are designed in such a way that uniform distribution of wastewater is always maintained.

The openings of orifices or nozzles on the laterals can be calculated as follows:

$$Q = 19.63\ Cd_1^2(h)^{0.5}\left[\frac{1}{1-\left(\frac{d_1}{d_2}\right)^4}\right]^{0.5} \quad \text{for } \frac{d_1}{d_2} > 0.3 \tag{45}$$

$$Q = 19.63\ Cd_1^2(h)^{0.5} \quad \text{for } \frac{d_1}{d_2} < 0.3 \tag{46}$$

where Q is the flow, gpm; d_1 is the diameter of orifice or nozzle opening, inches; h is the differential head at orifice, ft of liquid; d_2 is the diameter of lateral, inches; and C is the discharge coefficient.

Pipe lateral systems with orifices or nozzles of different sizes should be employed to insure that Q and h values are relatively constant at all orifices or nozzles.

4.2 Minimum Fluidization Velocity

The superficial upflow velocity (U) to be employed under a given set of operating conditions can be calculated via process kinetic equations developed in previous sections. Nevertheless, it is essential that, during the start-up of an FBBR, the superficial upflow velocity is always maintained above the minimum fluidization velocity of the media used. This practice provides a lower shear environment in which media exist and thus enhances the development of biofilms on the fluidized media. The minimum fluidization velocity (U_{mf}) can be calculated via the correlation developed by Wen and Yu [65]:

$$U_{mf} = \frac{0.5\,\mu}{\varrho_l r_m}\left[(33.7)^2 + 0.326\,\frac{\varrho_l r_m^3(\varrho_m - \varrho_l)\,g}{\mu^2}\right]^{0.5} - 33.7 \tag{47}$$

4.3 Control of Expanded Bed Height

Control of the expanded FBBR bed height, via biomass wasting and separation, provides the most direct and convenient means to maintain the equilibrium biofilm thickness at a desirable level. In practice, the expanded bed height is not constantly controlled at a given level by continuously wasting overgrown bioparticles from the reactor. If it were, the separated biomass would be carried away as a dilute sludge requiring further thickening before final disposal. This inevitably would increase the costs associated with sludge disposal.

A more practical way of controlling the expanded bed height is to allow it to fluctuate over a range determined by the desirable expanded bed height and the expansion rate of the fluidized bed. Assume that, under a given set of operating conditions, the desirable expanded bed height is set at H_B m and the expansion rate of the fluidized bed is λ m d^{-1}, then the fluidized bed is allowed to expand from $(H_B - \lambda/2)$m to $(H_B + \lambda/2)$m between biomass wasting and separation cycles. Sludge level detectors can be installed at these two elevations to detect the presence of the liquid-fluidized bed interface. Once the expanded bed height reaches the upper elevation, the separation pump is turned on and bioparticles are continuously pumped out of the reactor until the expanded bed height recesses to the lower elevation. Then the separation pump is turned off and the cycle of bed height control is completed. The desirable expanded bed height (H_B) can be calculated via the process kinetic equations developed in previous sections, and the expansion rate of the fluidized bed using the algorithm illustrated in Fig. 15.

4.4 Biofilm Separation and Biomass Wastage

In order to waste the required quantity of biomass from an FBBR to prevent excessive bed expansion it is necessary to separate the biomass from the media in a controlled fashion. Because the treatment and disposal of waste sludge account for a large fraction of the overall treatment cost, it is essential that waste sludge be produced directly with a high solids content. Several methods of biofilm separation and biomass wastage have been investigated for FBBRs. All of them were designed on the basis of mechanical separation of biofilm from media.

Cooper, et al. [70] of the Water Research Center (WRC), England, reported a new process for biofilm separation. In the WRC process, bioparticles are gently drained by gravity from the reactor into a shear tank. The bioparticles are allowed to settle for a few minutes in the shear tank before decanting the clear supernatant from the consolidated bioparticles. Following the decanting stage a high-speed mixer is turned on to shear the biomass from the media. Then the media/biomass slurry is fed to a vibrating sieve. The partially-cleaned media are retained on the deck of the vibrating sieve and returned to the reactor whereas the separated biomass passes through the sieve and then to the sludge storage tank for further disposal. For the denitrification FBBR operating with biomass concentrations in the range 15,000 to 30,000 mg VS L^{-1}, the WRC process can produce a waste sludge of about 8% dry solids.

Vibrating sieves similar to that reported by Cooper, et al. [70] have been used as a media/biomass separation means in various pilot-scale FBBR studies [5,7,8]. Bioparticles are pumped out of the reactor and fed directly onto the vibrating screen where the vibrational force separates the biofilms from the media. The cleaned media are returned to the reactor and the separated biomass is wasted as excess sludge.

A gravity fed stationary screen has been used as the media/biomass separation unit in a full-scale anaerobic FBBR [71]. The device employs a continuously curved bar screen 160 cm in length. A weir at the top of the device is utilized to tangentially introduce the feed flow (bioparticles and liquid) at right angles to the openings between the wedge bars of the concave screen surface. Partially-cleaned media are retained on the screen surface as they move downward under gravity whereas liquid and separated biomass are collected under the screen surface and carried into a side discharge outlet. No moving parts are involved in the operation of gravity fed stationary screens.

Other devices that have been investigated as a means for biofilm separation in FBBRs include the air-lift pump [70], helical screw conveyor [7], swirl concentrator/sand pump [4,5] and compression roller [72]. The compression roller is exclusively employed in the CAPTOR™ process in which small blocks of reticulated plastic foam are used as the media.

4.5 Recirculation Requirements

Mulcahy and LaMotta [43)] reported that, for typical nitrified municipal wastewaters, a denitrification FBBR can be operated as a plug-flow reactor without any provision for recirculation. Nevertheless, recirculation of reactor effluent is required in other applications for a variety of reasons.

When FBBRs are employed for carbon oxidation, nitrification and combined carbon oxidation-nitrification, recirculation of reactor effluent is required because of the low solubility of oxygen in wastewater. Unlike the conventional activated sludge processes when air or pure oxygen is added directly to the mixed liquor, in an FBBR pure oxygen is added to the wastewater before it enters the reactor. This oxygenation practice is employed to avoid the sloughing of biofilms from the media by rising gas bubbles. Thus, recirculation of reactor effluent is employed to insure that oxygen will not become a limiting factor.

Sutton, et al. [8)] reported that, in municipal applications of FBBRs about 60 mg L^{-1} O_2 can be dissolved in wastewater and then utilized in the reactor with less than 1% loss through the system when high purity oxygen is used as the supply. Effervescence in the reactor is negligible at this oxygen concentration. Thus, 58 mg L^{-1} of O_2 is consumed by biofilms in the reactor when effluent DO is 2 mg L^{-1}.

The recirculation requirement for a specific application can be calculated on the assumption that 1.0 kg of O_2 is required per kg BOD_5 removed plus 4.3 kg of O_2 required per kg of $NO_3^- - N$ produced:

$$r' = \frac{(S_0 - S_e) - 58}{58} \qquad \text{(carbon oxidation only)} \tag{48}$$

$$r' = \frac{(N_0 - N_e) - \left(\frac{58}{4.3}\right)}{\left(\frac{58}{4.3}\right)} \qquad \text{(nitrification only)} \tag{49}$$

$$r' = \frac{S_0 + 4.3\,N_0 - 58}{58 - S_e - 4.3\,N_e} \qquad \text{(carbon oxidation-nitrification)} \tag{50}$$

where r' is the recycle ratio; S_0 is the influent BOD_5, ML^{-3}; S_e is the effluent BOD_5, ML^{-3}; N_0 is the influent TKN, ML^{-3}; and N_e is the effluent TKN, ML^{-3}.

When FBBRs are employed for anaerobic treatment of industrial wastewaters, an adequate recycle ratio should be used to insure that the substrate loading rate does not exceed the design value as calculated using the process kinetic equations developed in previous sections.

4.6 Addition of Oxygen and Chemicals

High purity liquid oxygen is generally used in the aerobic FBBRs to maximize the transfer and utilization of oxygen in the reactor. To avoid excessive sloughing of biofilms from the fluidized media by rising gas bubbles, oxygen is added and dissolved into the wastewater before it enters the reactor via an oxygenation device operated under high pressure. Sutton, et al. [8] reported that, when high purity oxygen is used, about 60 mg L^{-1} of O_2 can be dissolved in an oxygenation device operated at a pressure of 2 atm (gauge) and a hydraulic retention time of about 25 s without significant effervescence in the reactor.

High purity oxygen can be obtained from on-site facilities. The constituent parts of such facilities would depend on the oxygen demand and the variations in wastewater flow rate. For low levels of oxygen demand, liquid oxygen can be supplied from a vacuum-insulated storage tank and a vaporization system. For intermediate levels of oxygen demand, liquid oxygen can be supplied from a pressure swing adsorption (PSA) plant, with peak demand supplied from liquid oxygen storage tank. For high levels of oxygen demand, a cryogenic oxygen plant together with a liquid storage tank can be used.

To accommodate diurnal variations in oxygen demand, the rate of oxygen addition is continuously monitored and controlled via a feedback control system which is designed to maintain a constant effluent DO. A DO measurement device located at the effluent end of the reactor is required to provide input information to the feedback control system. This practice provides nearly instantaneous control actions which significantly reduce operation and maintenance costs by lowering oxygen needs, reducing the amount of operator attention required, and maximizing the oxygenation efficiency.

Chemical addition is required when the alkalinity and/or nutrient content of the wastewater is insufficient to sustain the desirable substrate conversion rate in the FBBR. Chemicals in either dry powder or liquid form can be added directly into the recycle tank to provide desirable concentrations. Similar feedback control systems can be utilized for chemical feeding.

5 Current Applications

A large number and variety of wastewaters have been investigated on both the laboratory and pilot-scale using the FBBR technology. The applications can be categorized as municipal applications and industrial applications.

In municipal applications, FBBR's have been evaluated for all the conventional treatment processes: carbon oxidation, nitrification, denitrification, and anaerobic treatment [6]. For oxygenic FBBRs a volumetric loading rate of about 3 kg BOD_5 m^{-3} d^{-1} can be applied to produce effluent values of 20 mg L^{-1} BOD_5 and 30 mg L^{-1} SS. This value compares favorably with a design value of approximately 0.45 kg BOD_5 m^{-3} d^{-1} for conventional air activated sludge processes in the USA [8].

The U.S. Environmental Protection Agency (EPA) has awarded a grant to Nassau County, New York, to design and construct a 37,850 m^3 d^{-1} FBBR facility in addition to the existing 227,710 m^3 d^{-1} treatment plant [73]. The proposed FBBR facility consists of eight 5 m square by 5.18 m deep reactors, operating at a hydraulic loading rate of 36.7 m h^{-1}. Oxygen is supplied by a 5443 kg d^{-1} PSA plant and is dissolved in the wastewater via four 15.2 m deep U-tube aerators. Four 1.8 m diameter vibrating screens are to be provided for biofilm separation. Another full-scale oxygenic FBBR designed for treatment of combined municipal and canning wastewaters is to be under construction in California in near future.

Biological denitrification via anoxic FBBRs represents another promising municipal application of the FBBR technology. Four denitrification FBBRs, each 5.8 m × 5.8 m in cross section, have been constructed at the Advanced Waste Treatment Facility, Pensacola, Florida, USA [3]. However, no operating data are available for further evaluation.

Biological denitrification of river water used for potable water supply [11, 74, 75] and anaerobic treatment of municipal wastewaters [76] represent two potential municipal applications of the FBBR technology which are recently receiving widespread attention.

In industrial applications, FBBRs have been tested on laboratory and pilot scale for a variety of wastewaters. Anoxic FBBRs have been used to remove nitrates from explosive factory, petrochemical, nuclear industry [6, 10], and coke plant wastewaters [77]. Anaerobic FBBRs have been reported for the treatment of paper and pulp [19], molasses [78], food processing (corn chip) [79], chemical [79], soft drink bottling [79] and dairy wastewaters [18, 79, 80]. A full-scale anaerobic FBBR is under operation for treatment of soy processing wastewater [80]. This FBBR facility consists of four 6.1 m diameter by 12.5 m high reactors with a design capability of removing 8165 kg BOD_5 per day. The anaerobic FBBRs were designed as single-phase units. The temperature and pH of the reactors are controlled at 35 °C and 7.0 ± 0.2, respectively. To date, it is reported that this FBBR facility is operating with anticipated efficacy.

6 Future Outlook

There is potentially a very large market for FBBR technology in municipal applications since many existing treatment plants will need to be expanded and/or upgraded to meet more stringent standards in the future. The compact nature of FBBRs may allow upgrading within existing structures. Complete skid mounted systems can be directly shipped to site, requiring only piping and electrical connections to make them fully operational. Thus, existing plant operation will not be seriously interrupted and plant expansion and/or upgrading will be affected rapidly and economically. For larger installations, a portion of the existing tankage for suspended growth processes may be converted to the fluidized bed mode. This could be used as a quick, economical expansion alternative for treatment systems since the time for excavation and concrete form work is minimized and system down-time and capital cost expenditures are proportionally reduced. For installations which will have increasing hydraulic and organic loadings over a long period of time, the FBBR technology is particularly attractive since modules may be added as required or as effluent standards become more stringent.

Both the oxygenic and anoxic FBBRs are likely to be used in increasing numbers for upgrading existing municipal wastewater treatment plants.

The anaerobic FBBRs are already beginning to find a huge market for the treatment of high strength industrial wastewaters with COD concentration $> 1000 \text{ mg L}^{-1}$, because the high costs of oxygenation can be avoided and a useful gaseous fuel (i.e., methane) can be recovered. Sludge transport and disposal costs are lower than they are in the the oxygenic processes because of the production of less waste sludge under anaerobic conditions. The low capital and operating costs for anaerobic fluidized bed treatment provide an economical alternative to sewer surcharges and industrial cost recovery charges.

The FBBRs may also find many applications other than water and wastewater treatment, especially in commercial biomass conversion, ethanol and biochemical production, and microbial culturing where high reactor biomass holdup and the ability to harvest a very concentrated product are essential. The process intensification (i.e., the reduction in process size while maintaining performance) achieved in the FBBR makes this technology particularly well suited for these applications.

7 Notation

A	= Cross-sectional area of the reactor, L^2
C	= Discharge coefficient of the orifice or nozzle
C_D	= Drag coefficient
d_1	= Diameter of orifice or nozzle opening, inches
d_2	= Diameter of lateral, inches
d_m	= Media diameter, L
d_p	= Bioparticle diameter, L
D	= Diffusivity of substrate in water, L^2T^{-1}
D_e	= Effective diffusivity of substrate in biofilm, L^2T^{-1}
g	= Gravitational acceleration, LT^{-2}
h	= Differential head at the orifice or nozzle, ft of liquid
H_B	= Expanded bed height, L
$\hat{H}_B$	= Estimated expanded bed height, L
ΔH_B	= Change in expanded bed height, L
k_0	= Intrinsic zero order rate constant, $MM^{-1}T^{-1}$
k_1	= Intrinsic first order rate constant, $L^3M^{-1}T^{-1}$
k_c	= External mass transfer coefficient, LT^{-1}
K_0	= Observed zero order rate constant, $M^{0.5}L^{-2.5}$
K_1	= Observed first order rate constant, L^{-1}
n	= Expansion index
N_e	= Effluent TKN, ML^{-3}
N_0	= Influent TKN, ML^{-3}
N_{Re_t}	= Terminal Reynolds number
P	= Biofilm moisture content
Q	= Flow, gpm
r	= Radial distance measured from bioparticle center, L
r′	= Recycle ratio
$\bar{r}$	= Characteristic radius, L
r_c	= Substrate penetration depth, L
r_m	= Media radius, L
r_p	= Bioparticle radius, L
R	= observed substrate conversion rate per unit biofilm mass, $MM^{-1}T^{-1}$
R_v	= observed substrate conversion rate per unit fluidized bed volume, $ML^{-3}T^{-1}$
S	= intrabiofilm substrate concentration, ML^{-3}

S_b = Bulk-liquid substrate concentration, ML^{-3}
S_{bi} = Inlet substrate concentration, ML^{-3}
S_e = Effluent BOD_5, ML^{-3}
S_0 = Influent BOD_5, ML^{-3}
t = time, T
U = Superficial upflow velocity, LT^{-1}
U_{mf} = Minimum fluidization velocity, LT^{-1}
U_t = Bioparticle terminal settling velocity, LT^{-1}
V_m = Media volume, L^3
ΔV_m = Media volume in ΔH_B, L^3
V_s = Bioparticle volume, L^3
$W|_{r=r_p}$ = Mass transfer of substrate across liquid-biofilm interface, MT^{-1}
x = Expansion rate of bed height, m d^{-1}
X = Biomass concentration in an FBBR, ML^{-3}
X_a = Effective biomass concentration in an FBBR, ML^{-3}
Z = Axial position in an FBBR, L

Greek

∂ = $(\varrho k_1/D_e)^{0.5}$
ε = bed porosity
δ = biofilm thickness, L
δ = estimated biofilm thickness, L
ϱ = biofilm dry density, ML^{-3}
ϱ_l = liquid density, ML^{-3}
ϱ_m = media density, ML^{-3}
ϱ_s = bioparticle density, ML^{-3}
μ = liquid viscosity, MLT^{-2}
θ = hydraulic retention time, T
η = effectiveness factor
η_0 = bioparticle zero order effectiveness factor
η_1 = bioparticle first order effectiveness factor
Φ_0 = conventional zero order Thiele modulus
Φ_{0m} = modified zero order Thiele modulus
Φ_1 = conventional first order Thiele modulus
Φ_{1m} = modified first order Thiele modulus

8 References

1. Bailey, J. E., Ollis, D. F.: Biochemical Engineering Fundamentals, New York, McGraw-Hill 1977
2. Trawinski, H.: Chem. Ing.-Tech. *24*, 444 (1952)
3. Barbara, M., Flood, F., Jeris, J. S.: Clearwaters *6* (1979)
4. Jeris, J. S., Owen, R. W.: J. Water Roll. Control Fed. *49*, 816 (1977)
5. Jeris, J. S., Qwen, R. W., Hickey, R. F., Flood, F.: Water Poll. Control Fed. *49*, 816 (1977)
6. Cooper, P. F.: The Chemical Engineer 373 (1981)
7. Nutt, S. G., Stephenson, J. P., Pries, J. H.: Aerobic Fluidized Bed Treatment of Municipal Wastewater for Organic Carbon Removal presented at the 52nd Annual Conf. of the Water Pollution Control Federation, Houston, Texas 1979

8. Sutton, P. M., Shieh, W. K., Kos, P., Dunning, P. R.: Dorr-Oliver's Oxitron System™ Fluidized-Bed Water and Wastewater Treatment Process. In: Biological Fluidized Bed Treatment of Water and Wastewater (eds. Cooper, P. F., Atkinson, B.), p. 285, Chichester, England, Ellis Horwood Limited 1981
9. Mulcahy, L. T., Shieh, W. K., LaMotta, E. J.: Prog. Wat. Tech. *12*, 143 (1980)
10. Hancher, C. W., Taylor, P. A., Napier, J. M.: Biotech. Bioeng. Symp. *8*, 361 (1978)
11. Gauntlett, R. B.: Removal of Ammonia and Nitrate in the Treatment as of Potable Water. In: Biological Fluidized Bed Treatment of Water and Wastewater (eds. Cooper, P. F., Atkinson, B.), p. 48, Chichester, London, Ellis Horwood Limited 1981
12. Hermanowicz, S. W., Ganczarzyk, J. J.: Biotech. Bioeng. *25*, 1321 (1983)
13. Cooper, P. F., Wheeldon, D. H. V.: Complete Treatment of Sewage in a Two-Fluidized Bed System. In: Biological Fluidized Bed Treatment of Water and Wastewater (eds. Cooper, P. F., Atkinson, B.), p. 121, Chichester, London, Ellis Horwood Limited 1981
14. Jewell, W. J., Switzenbaum, M. S., Morris, J. W.: J. Water Poll. Control Fed. *53*, 482 (1981)
15. Switzenbaum, M. S., Jewell, W. J.: ibid. *52*, 1963 (1980)
16. Bull, M. A., Sterritt, R. M., Lester, J. N.: J. Chem. Tech. Biotechnol. *33B*, 221 (1983)
17. Schraa, G., Jewell, W. J.: J. Water Poll. Control Fed. *56*, 3, 226 (1984)
18. Boening, P. H., Larsen, W. F.: Biotech. Bioeng. *24*, 2539 (1982)
19. Hakulinen, R., Salonen, M. S.: Process Biochemistry 18 (1982)
20. LaMotta, E. J., Hickey, R. F., Buydos, J. F.: J. Env. Eng. Div., ASCE, *108*, EE6, 1326 (1982)
21. Characklis, W. G.: Water Research *7*, 1113 (1973)
22. Zobell, C. E., Allen, E. C.: J. Bacteriology *29*, 239 (1935)
23. Zobell, C. E., Anderson, D. Q.: Biol. Bulletin *71*, 324 (1936)
24. Zobell, C. E.: J. Bacteriology *46*, 39 (1943)
25. Heukelekian, H., Heller, A.: ibid. *40*, 547 (1940)
26. Heukelekian, H., Crosby, E. S.: Sewage Ind. Wasters *28*, 1, 73 (1956)
27. Jannasch, H. W., J. Gen. Microbiol. *18*, 609 (1958)
28. Bott, T. R., Miller, P. C.: J. Chem. Tech. Biotechnol. *33B*, 177 (1983)
29. Bryers, J. D., Characklis, W. G.: Biotech. Bioeng. *24*, 2451 (1982)
30. Sanders, W. M.: Water Research *3*, 81 (1967)
31. Kornegay, B. H., Andrews, J. F.: J. Water Poll. Control Fed. *40*, R460 (1968)
32. Marshall, K. C.: The Effects of Surfaces on Microbial Activity. In: Water Pollution Microbiology, Vol. 2 (ed. Mitchell, R.), p. 51, John Wiley & Sons, New York 1978
33. Wilkinson, J. F.: Bact. Review *22*, 46 (1958)
34. Lamanna, C., Mallette, M. F.: Basic Microbiology, Williams & Wilkins, Baltimore 1965
35. Gaudy, E., Wolfe, R. S.: Appl. Microbiol. *10*, 200 (1962)
36. Maier, W. J.: Mass Transfer and Growth Kinetics on a Slime Layer: Simulation of Trickling Filter. Ph. D. Dissertation, Cornell Univ. 1966
37. Mulcahy, L. T., Shieh, W. K., LaMotta, E. J.: Biotech. Bioeng. *23*, 2403 (1981)
38. Atkinson, B., Fowler, H. W.: The Significance of Microbial Film in Fermenters. In: Advances in Biochemical Engineering/Biotechnology (ed. Fiechter, A.), Vol. 3, 221, Heidelberg, Springer 1974
39. Characklis, W. G.: Biotech. Bioeng. *23*, 1923 (1981)
40. Shieh, W. K., Sutton, P. M., Kos, P.: J. Water Poll. Control Fed. *53*, 11, 1574 (1981)
41. Grady, C. P. L., Jr.: Modeling of Biological Fixed Films — A State-of-the-Art Review. In: Proceeding of the First International Conference on Fixed-Film Biological Processes (eds. Wu, Y. C., Smith, E. D., Miller, R. D., Patken, E. J. O.), Vol. 1, 344, Univ. of Pittsburgh, Pittsburgh, Pa. 1982
42. Hoehn, R. C., Ray, A. D.: J. Water Poll. Control Fed. *45*, 2302 (1973)
43. Mulcahy, L. T., LaMotta, E. J.: Mathematical Model of the Fluidized Bed Biofilm Reactor, Report No. Env. E. 59-78-2, Department of Civil Engineering, Univ. of Massachusetts/Amherst 1978
44. Tomlinson, T. G., Snadden, D. H. M.: Air & Water Poll./Int. Journ. *10*, 865 (1966)
45. Heukelekian, H., Crosby, E. S.: Sewage Ind. Wastes *28*, 2, 206 (1956)
46. Metcalf & Eddy, Inc.: Wastewater Engineering: Treatment/Disposal/Reuse, 2nd Edition, New York, McGraw-Hill 1979
47. Paolini, A. E., Sebastianti, E., Variali, G.: Water Research *13*, 751 (1979)
48. Williamson, K. J., McCarty, P. L.: J. Water Poll. Control Fed. *48*, 28 (1976)

49. Bungay, H. R. IV, Whalen, W. J., Sanders, W. M.: Biotech. Bioeng. *11*, 765 (1969)
50. Chen, Y. S., Bungay, H. R. IV: ibid. *23*, 781 (1981)
51. LaMotta, E. J.: Evaluation of Diffusional Resistances in Substrate Utilization by Biological Films. Ph. D. Dissertation, University of North Carolina at Chapel Hill 1976
52. Atkinson, B., Daoud, I. S., Williams, D. A.: Trans. Instn. Chem. Engrs. *46*, T245 (1968)
53. Atkinson, B., How, S. Y.: ibid. *52*, 260 (1974)
54. Alleman, J. E., Veil, J. A., Canaday, J. T.: Water Research *16*, 543 (1982)
55. Friedman, B. A., Dugan, P. R.: J. Bacteriology *95*, 1903 (1968)
56. Heukelekian, H.: Sewage Ind. Wastes *28*, 78 (1956)
57. Jeris, J. S., Beer, C., Mueller, J. A.: J. Water Poll Control Fed. *47*, 2043 (1975)
58. Atkinson, B., Swilley, E. L.: Water Research *1*, 687 (1967)
59. Shieh, W. K., Mulcahy, L. T., LaMotta, E. J.: Trans. Instn. Chem. Engr. *59*, 129 (1981)
60. Aris R.: Elementary Chemical Reactor Analysis. Englewood Cliffs, N. J., Prentice-Hall 1969
61. Webb, C., Black, G. M., Atkinson, B.: Chem. Eng. Res. Des. *61*, 125 (1983)
62. Shieh, W. K., Chen, C. Y.: ibid. *62*, 133 (1984)
63. Richardson, J. F., Zaki, W. N.: Trans. Instn. Chem. Engr. *32*, 35 (1954)
64. Lewis, E. W., Bowerman, E. W.: Chem. Eng. Prog. *48*, 605 (1952)
65. Wen, C. Y., Yu, Y. H.: Chem. Eng. Prog. Symp. Series *62*, 100 (1962)
66. Vasalos, I. A., Rundell, D. N., Megiris, K. E., Tjatjopoulos, G. T.: AIChE J. *28*, 2, 346 (1982)
67. Harremoes, P.: Vatten *33*, 122 (1977)
68. Kobayashi, T., Laidler, K. J.: Biochemica et Biophysica Acta 302 (1973)
69. Mulcahy, L. T., Shieh, W. K., LaMotta, E. J.: Water-1980 AIChE Symp. Ser., *209*, 77, 273 (1981)
70. Cooper, P. F., Wheeldon, D. H. V., Ingram-Tedd, P. E., Harrington, D. W.: Sand/Biomass Separation with Production of a Concentrated Sludge. In: Biological Fluidized Bed Treatment of Water and Wastewater (eds. Cooper, P. F., Atkinson, B.), p. 361, Chichester, London, Ellis Horwood Limited 1981
71. Li, A.: Personal Communication (1984)
72. Simon-Hartley, Ltd.: Technical Bulletin (1984)
73. Oppelt, E. T., Smith, J. M.: United States Environmental Protection Agency Research and Current Thinking on Fluidized-Bed Biological Treatment. In: Biological Fluidized Bed Treatment of Water and Wastewater (eds. Cooper, P. F., Atkinson, B.), p. 165, Chichester, London, Ellis Horwood Limited 1980
74. Richard, Y., Leprince, A., Martin, G., LeBlanc, C.: Prog. Wat. Tech. *12*, 173 (1980)
75. Cooper, P. F., Wheeldon, D. H. V.: Wat. Poll. Control 286 (1980)
76. Switzerbaum, M. S.: Enzyme Microb. Technol. 243 (1983)
77. Nutt, S. G., Melcer, H., Pries, J. H.: J. Water Poll. Control Fed. *56*, 7, 851 (1984)
78. Frostell, B.: Process Biochemistry 37 (1982)
79. Hickey, R. F., Owens, R. W.: Biotech. Bioeng. Symp. *11*, 399 (1981)
80. Sutton, P. M., Li, A., Evans, R. R., Korchin, S.: Dorr-Oliver's Fixed Film and Suspended Growth Anaerobic Systems For Industrial Wastewater Treatment and Energy Recover. Presented at the 37th Annual Purdue Industrial Waste conf., West Lafayette, Indiana 1982

ISBN 3-05-500205-9

Bei Fragen zur Produktsicherheit wenden Sie sich bitte an:
If you have any questions regarding product safety,
please contact:

Walter de Gruyter GmbH
Genthiner Straße 13
10785 Berlin
productsafety@degruyterbrill.com